Extreme States of Matter

Joseph A. Angelo, Jr.

To all volunteers who search for and help recover lost dogs, cats, and other pets, especially those wonderful people who assisted in the miraculous recovery of little Mandy the Pug from a Florida wilderness area in early January 2011

EXTREME STATES OF MATTER

Facts On File, Inc.
An imprint of Infobase Learning
132 West 31st Street
New York NY 10001

Library of Congress Cataloging-in-Publication Data
Angelo, Joseph A.
Extreme states of matter / Joseph A. Angelo, Jr.
p. cm.—(States of matter)
Includes bibliographical references and index.
ISBN 978-0-8160-7606-2
1. Matter—Properties—Popular works. 2. Chemical elements—Popular works. 3. Exobiology—Popular works. I. Title.
QC171.2.A54 2012
530.4—dc22 2011006002

Facts On File books are available at special discounts when purchased in bulk quantities for businesses, associations, institutions, or sales promotions. Please call our Special Sales Department in New York at (212) 967-8800 or (800) 322-8755.

You can find Facts On File on the World Wide Web at http://www.infobaselearning.com

Text design by Annie O'Donnell
Composition by Hermitage Publishing Services
Illustrations by Sholto Ainslie
Photo research by the Author
Cover printed by IBT Global, Troy, N.Y.
Book printed and bound by IBT Global, Troy, N.Y.
Date printed: December 2011
Printed in the United States of America

10 9 8 7 6 5 4 3 2 1

This book is printed on acid-free paper.

Contents

Preface

The unleashed power of the atom has changed everything save our modes of thinking.

—Albert Einstein

Humankind's global civilization relies upon a family of advanced technologies that allow people to perform clever manipulations of matter and energy in a variety of interesting ways. Contemporary matter manipulations hold out the promise of a golden era for humankind—an era in which most people are free from the threat of such natural perils as thirst, starvation, and disease. But matter manipulations, if performed unwisely or improperly on a large scale, can also have an apocalyptic impact. History is filled with stories of ancient societies that collapsed because local material resources were overexploited or unwisely used. In the extreme, any similar follies by people on a global scale during this century could imperil not only the human species but all life on Earth.

Despite the importance of intelligent stewardship of Earth's resources, many people lack sufficient appreciation for how matter influences their daily lives. The overarching goal of States of Matter is to explain the important role matter plays throughout the entire domain of nature—both here on Earth and everywhere in the universe. The comprehensive multivolume set is designed to raise and answer intriguing questions and to help readers understand matter in all its interesting states and forms—from common to exotic, from abundant to scarce, from here on Earth to the fringes of the observable universe.

The subject of matter is filled with intriguing mysteries and paradoxes. Take two highly flammable gases, hydrogen (H_2) and oxygen (O_2), carefully combine them, add a spark, and suddenly an exothermic reaction takes place yielding not only energy but also an interesting new substance called water (H_2O). Water is an excellent substance to quench a fire, but it is also an incredibly intriguing material that is necessary for all life here on Earth—and probably elsewhere in the universe.

Matter is all around us and involves everything tangible a person sees, feels, and touches. The flow of water throughout Earth's biosphere, the air people breathe, and the ground they stand on are examples of the most commonly encountered states of matter. This daily personal encounter with matter in its liquid, gaseous, and solid states has intrigued human beings from the dawn of history. One early line of inquiry concerning the science of matter (that is, *matter science*) resulted in the classic earth, air, water, and fire elemental philosophy of the ancient Greeks. This early theory of matter trickled down through history and essentially ruled Western thought until the Scientific Revolution.

It was not until the late 16th century and the start of the Scientific Revolution that the true nature of matter and its relationship with energy began to emerge. People started to quantify the properties of matter and to discover a series of interesting relationships through carefully performed and well-documented experiments. Speculation, philosophical conjecture, and alchemy gave way to the scientific method, with its organized investigation of the material world and natural phenomena.

Collectively, the story of this magnificent intellectual unfolding represents one of the great cultural legacies in human history—comparable to the control of fire and the invention of the alphabet. The intellectual curiosity and hard work of the early scientists throughout the Scientific Revolution set the human race on a trajectory of discovery, a trajectory that not only enabled today's global civilization but also opened up the entire universe to understanding and exploration.

In a curious historical paradox, most early peoples, including the ancient Greeks, knew a number of fundamental facts about matter (in its solid, liquid, and gaseous states), but these same peoples generally made surprisingly little scientific progress toward unraveling matter's inner mysteries. The art of metallurgy, for example, was developed some 4,000 to 5,000 years ago on an essentially trial-and-error basis, thrusting early civilizations around the Mediterranean Sea into first the Bronze Age and later the Iron Age. Better weapons (such as metal swords and shields) were the primary social catalyst for technical progress, yet the periodic table of chemical elements (of which metals represent the majority of entries) was not envisioned until the 19th century.

Starting in the late 16th century, inquisitive individuals, such as the Italian scientist Galileo Galilei, performed careful observations and measurements to support more organized inquiries into the workings of the natural world. As a consequence of these observations and experiments, the nature of matter became better understood and better quantified. Scientists introduced the concepts of density, pressure, and temperature in their efforts to more consistently describe matter on a large (or macroscopic) scale. As instruments improved, scientists were able to make even better measurements, and soon matter became more clearly understood on both a macroscopic and microscopic scale. Starting in the 20th century, scientists began to observe and measure the long-hidden inner nature of matter on the atomic and subatomic scales.

Actually, intellectual inquiry into the microscopic nature of matter has its roots in ancient Greece. Not all ancient Greek philosophers were content with the prevailing earth-air-water-fire model of matter. About 450 B.C.E., a Greek philosopher named Leucippus and his more well-known student Democritus introduced the notion that all matter is actually composed of tiny solid particles, which are *atomos* (ατομος), or indivisible. Unfortunately, this brilliant insight into the natural order of things lay essentially unnoticed for centuries. In the early 1800s, a British schoolteacher named John Dalton began tinkering with mixtures of gases and made the daring assumption that a chemical element consisted of identical indestructible atoms. His efforts revived atomism. Several years later, the Italian scientist Amedeo Avogadro announced a remarkable hypothesis, a bold postulation that paved the way for the atomic theory of chemistry. Although this hypothesis was not widely accepted until the second half of the 19th century, it helped set the stage for the spectacular revolution in matter science that started as the 19th century rolled into the 20th.

What lay ahead was not just the development of an atomistic kinetic theory of matter, but the experimental discovery of electrons, radioactivity, the nuclear atom, protons, neutrons, and quarks. Not to be outdone by the nuclear scientists, who explored nature on the minutest scale, astrophysicists began describing exotic states of matter on the grandest of cosmic scales. The notion of degenerate matter appeared as well as the hypothesis that supermassive black holes lurked at the centers of most large galaxies after devouring the masses of millions of stars. Today, cosmologists and astrophysicists describe matter as being clumped into enormous clusters and superclusters of galaxies. The quest for these scientists is to explain how the observable universe, consisting of understandable forms of matter and energy, is also immersed in and influenced by mysterious forms of matter and energy, called dark matter and dark energy, respectively.

The study of matter stretches from prehistoric obsidian tools to contemporary research efforts in nanotechnology. States of Matter provides 9th- to 12th-grade audiences with an exciting and unparalleled adventure into the physical realm and applications of matter. This journey in search of the meaning of substance ranges from everyday "touch, feel, and see" items (such as steel, talc, concrete, water, and air) to the tiny, invisible atoms, molecules, and subatomic particles that govern the behavior and physical characteristics of every element, compound, and mixture, not only here on Earth, but everywhere in the universe.

Today, scientists recognize several other states of matter in addition to the solid, liquid, and gas states known to exist since ancient times. These include very hot plasmas and extremely cold Bose-Einstein condensates. Scientists also study very exotic forms of matter, such as liquid helium (which behaves as a superfluid does), superconductors, and quark-gluon plasmas. Astronomers and astrophysicists refer to degenerate matter when they discuss white dwarf stars and neutron stars. Other unusual forms of matter under investigation include antimatter and dark matter. Perhaps most challenging of all for scientists in this century is to grasp the true nature of dark energy and understand how it influences all matter in the universe. Using the national science education standards for 9th- to 12th-grade readers as an overarching guide, the States of Matter set provides a clear, carefully selected, well-integrated, and enjoyable treatment of these interesting concepts and topics.

The overall study of matter contains a significant amount of important scientific information that should attract a wide range of 9th- to 12th-grade readers. The broad subject of matter embraces essentially all fields of modern science and engineering, from aerodynamics and astronomy, to medicine and biology, to transportation and power generation, to the operation of Earth's amazing biosphere, to cosmology and the explosive start and evolution of the universe. Paying close attention to national science education standards and content guidelines, the author has prepared each book as a well-integrated, progressive treatment of one major aspect of this exciting and complex subject. Owing to the comprehensive coverage, full-color illustrations, and numerous informative sidebars, teachers will find the States of Matter to be of enormous value in supporting their science and mathematics curricula.

Specifically, States of Matter is a multivolume set that presents the discovery and use of matter and all its intriguing properties within the context of science as inquiry. For example, the reader will learn how the ideal gas law (sometimes called the ideal gas equation of state) did not happen overnight. Rather, it evolved slowly and was based on the inquisitiveness and careful observations of many scientists whose work spanned a period of about 100 years. Similarly, the ancient Greeks were puzzled by the electrostatic behavior of certain matter. However, it took several millennia until the quantified nature of electric charge was recognized. While Nobel Prize–winning British physicist Sir J. J. (Joseph John) Thomson was inquiring about the fundamental nature of electric charge in 1898, he discovered the first subatomic particle, which he called the electron. His work helped transform the understanding of matter and shaped the modern world. States of Matter contains numerous other examples of science as inquiry, examples strategically sprinkled throughout each volume to show how scientists used puzzling questions to guide their inquiries, design experiments, use available technology and mathematics to collect data, and then formulate hypotheses and models to explain these data.

States of Matter is a set that treats all aspects of physical science, including the structure of atoms, the structure and properties of matter, the nature of chemical reactions, the behavior of matter in motion and when forces are applied, the mass-energy conservation principle, the role of thermodynamic properties such as internal energy and entropy (disorder principle), and how matter and energy interact on various scales and levels in the physical universe.

The set also introduces readers to some of the more important solids in today's global civilization (such as carbon, concrete, coal, gold, copper, salt, aluminum, and iron). Likewise, important liquids (such as water, oil, blood, and milk) are treated. In addition to air (the most commonly encountered gas here on Earth), the reader will discover the unusual properties and interesting applications of other important gases, such as hydrogen, oxygen, carbon dioxide, nitrogen, xenon, krypton, and helium.

Each volume within the States of Matter set includes an index, an appendix with the latest version of the periodic table, a chronology of notable events, a glossary of significant terms and concepts, a helpful list of Internet resources, and an array of historical and current print sources for further research. Based on the current principles and standards in teaching mathematics and science, the States of Matter set is essential for readers who require information on all major topics in the science and application of matter.

Acknowledgments

I wish to thank the public information and/or multimedia specialists at the U.S. Department of Energy (including those at DOE headquarters and at all the national laboratories), the U.S. Department of Defense (including the individual armed services: U.S. Air Force, U.S. Army, U.S. Marines, and U.S. Navy), the National Institute of Standards and Technology (NIST) within the U.S. Department of Commerce, the U.S. Department of Agriculture, the National Aeronautics and Space Administration (NASA) (including its centers and astronomical observatory facilities), the National Oceanic and Atmospheric Administration (NOAA) of the U.S. Department of Commerce, and the U.S. Geological Survey (USGS) within the U.S. Department of the Interior for the generous supply of technical information and illustrations used in the preparation of this book set. Also recognized are the efforts of Frank K. Darmstadt and other members of the Facts On File team, whose careful attention to detail helped transform an interesting concept into a polished, publishable product. The continued support of two other special people must be mentioned here. The first individual is my longtime personal physician, Dr. Charles S. Stewart III, M.D., whose medical skills allowed me to successfully work on this interesting project. The second individual is my wife, Joan, who for the past 45 years has provided the loving and supportive home environment so essential for the successful completion of any undertaking in life.

Introduction

It is difficult to say what is impossible, for the dream of yesterday is the hope of today and the reality of tomorrow.

—Robert Hutchings Goddard (1882–1945)
American rocket scientist

The history of civilization is essentially the story of the human mind understanding matter. *Extreme States of Matter* takes the reader on an incredible journey across the most exciting scientific frontiers of the 21st century. These intellectual frontiers have emerged from the practice of science, primarily involve the scientific interpretation of substance, and correspond to areas of advanced research taking place around the world. Complementary activities in nuclear physics (involving minute, subatomic quantities of matter) and astrophysics (involving the behavior of matter on the cosmic scale) are providing scientists with exciting new insights into the nature of matter and energy.

Extreme States of Matter describes the unusual and almost bizarre characteristics and properties of matter at extreme states. Such extreme states include matter at exceptionally high temperatures, exceptionally low temperatures, incredibly high pressures, intense magnetic fields, and intense gravitational fields. Scientists are curious individuals who often like to speculate about nature under conditions well beyond the conditions encountered and measured in their laboratories here on Earth. For example, unlike problems in classical mechanics, where time is assumed a constant, in Albert Einstein's special relativity, time becomes an interesting variable whenever matter experiences velocities approaching the speed of light.

This volume takes the reader on a fascinating journey into the realm of frontier science, where speculations about the extreme states of matter have promoted great progress in the past and are about to do the same in this century. The three most familiar states of matter found on Earth's surface are solid, liquid, and gas. When temperatures get sufficiently high, another state of matter called plasma appears. Finally, as temperatures

Stars are the fundamental collections of mass in the observable universe. The *Hubble Space Telescope* image shows about 100,000 stars in the globular cluster found in the constellation Aquarius. Astronomers refer to this globular cluster as Messier (M) 72 or New General Catalog (NGC) 6981. M72 is about 50 light-years in diameter and lies more than 50,000 light-years away from Earth. *(NASA/ESA/STScI)*

get very low and approach absolute zero, scientists encounter still another state of matter called the Bose-Einstein condensate (BEC).

Some of the major topics appearing in this book include: plasmas (very hot matter), Bose-Einstein condensates (extremely cold matter), degenerate matter (white dwarfs and neutron stars), black holes, antimatter, dark matter, and dark energy. There are even more speculative topics, such as the concept of a wormhole, the possibility of parallel universes, and the ultimate fate of matter in the current universe. When viewed against all the observable matter present in the universe, living matter represents a very special, highly evolved and complex state of substance. This book briefly explores the nature of thinking matter. The physics of consciousness represents the interesting physical state when certain forms of matter become self-aware and cognizant.

Scientists define plasma as an electrically neutral gaseous mixture of positive and negative ions. Stars are the basic unit of matter in the observable universe. Many readers will be surprised to learn that plasmas are the most common form of (ordinary) matter—comprising more than 99 percent of the visible universe. Plasmas permeate the solar system, as well as interstellar and intergalactic environments. Later in this century, an ability to manipulate, sustain, and control high-density plasmas could lead to the successful application of controlled thermonuclear fusion power here on Earth.

Extreme States of Matter shows how science at the limits of knowledge involves people and history. Some of today's most incredible frontier ideas actually appeared decades or even centuries ago, when curious people began thinking about matter in some extreme state or condition. For example, scientists trace the basic concept of the atomic structure of matter (that is, atomism) back to ancient Greece. In the fifth century B.C.E., the Greek philosophers Leucippus and his famous pupil Democritus introduced the theory of atomism within an ancient society that was more comfortable assuming that four basic elements, earth, air, water, and fire, made up all the matter in the world. As a natural philosopher but not an experimenter, Democritus used his mind to consider what would happen if he kept cutting a piece of matter, any type matter, into finer and finer halves. He reasoned that he would eventually reach the point where any further division would be impossible. As a result, the modern word *atom* comes from the ancient Greek word *atomos* (ατομος), which means "not divisible." In the 21st century, scientists use large, powerful particle accelerators to explore quark-gluon plasmas (QGPs) in an effort to unlock additional secrets from nature's tiniest pieces of matter. They also employ computer-generated virtual reality environments to examine physical reality in varying degrees of detail.

This book describes how scientists developed the quantum-level model of matter, which they refer to as the standard model. The comprehensive model explains reasonably well what the material world consists of and how it holds itself together. Within this model, physicists need only six quarks and six leptons to explain matter. The only fundamental particle predicted by the standard model that has not yet been observed is a hypothetical particle called the Higgs boson. The British theoretical physicist Peter Higgs hypothesized in 1964 that this type of particle may explain why certain elementary particles (such as quarks and electrons) have mass and other particles (such as photons) do not. If discovered by research scientists in this century, the Higgs boson (sometimes called the God particle) could play a major role in refining the standard model and shed additional light on the nature of matter at the quantum level.

However, the standard model does not explain everything of interest to scientists. One obvious omission is the fact that the standard model does not include gravity. *Extreme States of Matter* also presents some the latest ideas involving quantum gravity. In 1784, the British geologist and natural philosopher John Michell mentally pushed Sir Isaac Newton's universal law of gravitation to the extreme when he imagined an object so massive that it would prevent light from escaping. Michell called this postulated object a dark star. His intuitive speculations anticipated the refinement of the black hole concept by 20th century physicists. As described in the book, black hole physics represents one of the most exciting frontiers in contemporary science.

Similarly, the German astronomer Heinrich Wilhelm Olbers formulated (in about 1826) the famous philosophical discussion that became known as Olbers's paradox. Assuming a universe of infinite size and containing an infinite number of stars, he logically asked the question: Why is the night sky not as bright as the surface of the Sun? The contemporary response to Olbers's intriguing question relies on 20th-century developments in cosmology, including the concept of an expanding universe and the idea of the big bang.

The big bang is a widely accepted theory in contemporary cosmology concerning the origin and evolution of the universe. According to the big bang cosmological model, about 13.7 billion years ago there was an incredibly powerful explosion that started the present universe. Before

(opposite page) A scientist at the Argonne National Laboratory (ANL) uses an automatic virtual reality environment to examine the way complex biological molecules link up and form strings. The scientist can stop the simulation at any point and quite literally move around inside the image of this Lilliputian world to see changes and unique linkages from all angles. This application of virtual reality is a powerful starting point in nanotechnology-level biomedical research, pharmaceutical development projects, and materials science. *(DOE/ANL)*

this ancient explosion, matter, energy, space, and time did not exist. All of these physical phenomena emerged from an unimaginably small, infinitely dense object called the initial singularity. Immediately after the big

bang, the intensely hot universe, which consisted entirely of pure energy, began to expand and cool. As the universe cooled, matter began to form and undergo a variety of interesting transformations.

Recent astrophysical observations indicate that the universe has been expanding ever since, but at varying rates of acceleration due to a continual cosmic tug of war between the mysterious pushing force of dark energy and the gravitational pulling force of matter—both ordinary matter and dark matter. *Extreme States of Matter* summarizes the latest scientific work concerning dark matter and dark energy. Scientists define dark matter as matter in the universe that cannot be observed directly because it emits very little or no electromagnetic radiation and experiences little or no directly measurable interaction with ordinary matter. They currently think that the most common substance in the universe is dark matter—exceeding the total amount of ordinary (or baryonic) matter by a factor of five or more.

Near the end of the 20th century, big bang cosmologists had barely grown comfortable with a gravity-dominated, expanding universe model, when another major cosmological surprise popped up. In 1998, two separate teams of astrophysicists independently observed that the universe was not just expanding, but expanding *at an increasing rate.* What was causing this to happen? Since that startling discovery, scientists have attempted to explain the increasing rate of expansion of the universe by endorsing the notion of a special force called dark energy, which opposes gravity and allows the universe to expand at an ever more rapid rate.

Extreme States of Matter also discusses antimatter and degenerate matter. When a particle meets its antiparticle, they annihilate each other, and their combined annihilating masses reappear in the form of pure energy. Degenerate matter represents extremely high density matter. A degenerate star, such as a white dwarf or a neutron star, is one that has collapsed under the influence of its own gravity and formed an incredibly dense celestial object.

The ability of human beings to relate the microscopic (atomic level) behavior of matter to readily observable macroscopic properties (such as density, pressure, and temperature) helped transform previous civilizations. Breakthroughs in such areas as black hole physics, an understanding of dark matter, and a scientific interpretation of dark energy could propel the human race on a trajectory of technical development at a rate unprecedented in all previous history.

Extreme States of Matter examines superconductivity—a frontier area within one of matter's most interesting and important properties, electromagnetism. Scientists define superconductivity as the ability of a material to conduct electricity without resistance. Readers will discover that scientists call the temperature at which a material suddenly experiences zero (or negligible) electrical resistivity as that material's critical temperature or the transition temperature. When cooled below its critical temperature, a material becomes a superconductor. In 1911, the Dutch physicist Heike Onnes observed that the electrical resistance in a test sample of mercury (Hg) essentially vanished at –452.11°F (–268.95°C [4.2 K]).

In the 1980s, the German physicist Johannes Georg Bednorz and the Swiss physicist Karl Alexander Müller started a systematic investigation of the electrical properties of certain ceramic materials. By 1986, they succeeded in demonstrating that lanthanum barium copper oxide (LaBaCuO, or LBCO) exhibited superconductivity at a critical temperature of –396.7°F (–238.2°C [35 K]). This was an important milestone in the development of high-temperature superconductor materials. For their discovery of superconductivity in ceramic materials, Bednorz and Müller shared the 1987 Nobel Prize in physics. As suggested in the book, ceramic materials that function at or above liquid nitrogen temperatures significantly could expand the beneficial applications of superconductivity. Compared to liquid helium, liquid nitrogen is a more easily handled and much less expensive coolant.

Extreme States of Matter has been carefully designed to help any student or teacher who has an interest in the unusual states of matter discover how scientists measure and characterize these extreme states. Readers will also learn how the fascinating properties and characteristics of extreme-state matter might influence the course of human civilization in this century. The concluding portion of the book contains an appendix with a contemporary periodic table, chronology, glossary, and an array of historical and current sources for further research. These should prove especially helpful for readers who need additional information on specific terms, topics, and events.

The author has developed *Extreme States of Matter* so that a person familiar with SI units (the international language of science) will have no difficulty understanding and enjoying its contents. The author also recognizes that there is a continuing need for some students and teachers in the United States to have units expressed in the American (United States) customary system of units. Wherever appropriate, both unit systems appear side by side. An editorial decision places the American customary units first, followed by the equivalent SI units in parentheses. This format does not imply the author's preference for American customary units over SI units. Rather, the author strongly encourages all readers to take advantage of the particular formatting arrangement to learn more about the important role that SI units play within the international scientific community.

CHAPTER 1

An Initial Look at Matter Nearing Extreme Conditions

This chapter introduces the basic macroscopic properties of *mass, density, pressure,* and *temperature.* It also explains how scientists use these properties to describe the common physical states and general behavior of matter on Earth. Several examples of unusual values of these fundamental properties help to illustrate the behavior of matter approaching extreme conditions on Earth, as well as elsewhere in the universe. The chapter concludes with a discussion of time, the important role it plays in science, and the interesting concept of the arrow of time.

MASS: THE MEASUREMENT OF SUBSTANCE

Scientists define matter as anything that has mass and occupies space. Matter is the substance of all material objects in the universe. The three most common forms of matter found on Earth are *solid, liquid,* and *gas.* Subsequent chapters of this book describe two other interesting states of matter: *plasma* and the *Bose-Einstein condensate* (BEC).

One important factor shared by all forms of matter is that mass gives rise to the important physical property called *inertia.* Inertia is the resistance of a body to a change in its state of motion. As first proposed by Sir Isaac Newton (1642–1727) in the 17th century, mass is the inherent property of a body that helps scientists quantify inertia.

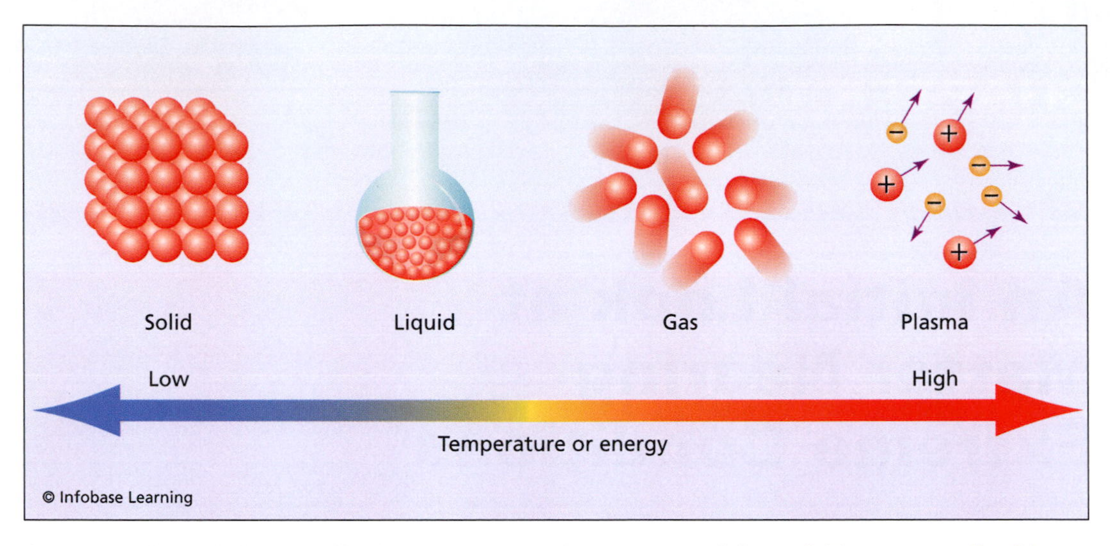

As energy is added to a solid, its temperature increases and the solid becomes a liquid. Further addition of energy allows the liquid to become a gas. If the gas is heated a great deal more, its atoms break apart into charged particles (that is, electrons and positively charged nuclei), resulting in the fourth state of matter called plasma. *(Based on NASA-sponsored artwork)*

When examined at the macroscopic level, solid matter retains its shape, occupies a specific volume, and exhibits important bulk physical properties such as density, ductility, *elasticity,* strength, and hardness. When viewed at the microscopic (or atomic) level, the *atoms* or *molecules* of a solid are not very mobile, and have the tendency to stay relatively fixed in their positions with respect to adjacent atoms or molecules.

The addition of *heat* can cause a solid to melt and experience a *change of state,* becoming a liquid. (The expression *phase change* is also commonly encountered during discussions of melting or *evaporation.*) On the macroscopic scale, the liquid form of a substance is quite different from the solid form of that substance.

A liquid will assume the shape of its container and will occupy a specific volume within that container. When viewed at the atomic (or microscopic) level, the atoms or molecules that make up a liquid are free to move about much more than they can in a solid. Resembling miniature versions of shape-shifter creatures found in science fiction, the atoms or molecules of a liquid can slip and slide past one another and assume the shape of their container. But the *forces* between atoms or molecules in a

MASS–ENERGY

Nuclear weapons have indelibly impressed upon the human psyche the equivalence of mass and energy, as prescribed by Einstein's special relativity theory. Shown here is the atmospheric detonation of a 15-kiloton (kT)-yield nuclear device (code-named Grable) at the Nevada Test Site on May 25, 1953. *(DOE/NTS)*

In 1905, Albert Einstein (1879–1955) wrote "On the Electrodynamics of Moving Bodies." This paper introduced his special theory of relativity, which deals with the laws of physics as seen by observers moving relative to one another at constant *velocity.* Einstein stated the first postulate of special relativity as: "The speed of light *(c)* has the same value for all (inertial-reference frame) observers, independent and regardless of the motion of the light source or the observers." His second postulate of special relativity proclaimed: "All physical laws are the same for all observers moving at constant *velocity* with respect to each other."

From the special theory of relativity, Einstein concluded that only a zero rest mass particle, such as a *photon,* could travel at the *speed of light.* A major consequence of special relativity is the equivalence of mass and *energy.* Einstein's famous mass-energy formula, expressed as $E = mc^2$, provides the energy equivalent of matter and vice versa. Among its many important physical insights, this equation was the key that scientists needed to understand energy release in such important nuclear reactions as *fission, fusion,* radioactive decay, and matter-antimatter annihilation.

(continues)

(continued)

An electron has a mass of approximately 2.01 × 10^{-30} lbm (9.11 × 10^{-31} kg). The energy equivalent of an electron at rest is about 511 keV, which corresponds to 7.76 × 10^{-17} *British thermal unit* (Btu) (8.19 × 10^{-14} *joules* [J]). As a point of reference, an *electron volt* (eV) has the energy equivalent of 1.519 × 10^{-22} Btu (1.60 × 10^{-19} J). Now consider the mass-energy of something much larger than an electron, such as a U.S. one cent coin. A typical American one cent coin has a mass of about 6.62 × 10^{-3} lbm (3 × 10^{-3} kg [3 g]). If all the matter in just one penny is suddenly converted into pure energy, the amount of energy released would be quite large, about 2.56 × 10^{11} Btu (2.7 × 10^{14} J).

liquid still exert considerable influence—that is why a liquid substance maintains a distinct, observable *volume.*

If water in a flask is carefully heated, it will eventually boil and evaporate into another common form of matter called a gas. (Scientists often use the term *vapor* to describe a gas that readily condenses back to the liquid state). At the macroscopic level, gases (including water vapor) do not keep their shape and do not maintain any definite volume. As a result, gases will simply occupy all the available space in a closed container. Gases will escape from an open container into the surrounding (lower pressure) environment until a uniform pressure condition, characterized by mechanical equilibrium, is achieved.

At the microscopic level, the atoms or molecules of a gas do not experience any significant interatomic or intermolecular forces. Rather, the atoms or molecules of a gas can travel relatively unhindered at high speed throughout the entire enclosing volume until they physically collide with one another or the container wall. In the 19th century, scientists discovered that the average speed of atoms or molecules in a gas is related to the absolute temperature of the gas. Specifically, scientists observed that the higher the temperature of a gas, the higher the average speed of its constituent atoms or molecules.

Physicists call solids and liquids condensed matter and gases uncondensed matter. They refer to the extremely dense matter found in white dwarf stars and neutron stars as degenerate matter. *Antimatter* consists of fundamental particles that have the opposite electric charges of ordinary matter—namely, negative *protons* and positive *electrons*. (Although the antineutron is uncharged, the antiparticle has the opposite magnetic spin and therefore experiences annihilation when it encounters an ordinary neutron.) Finally, although making up about 80 percent of the total mass of the universe, scientists cannot directly observe *dark matter* because it emits very little or no electromagnetic radiation.

Mass (m) is the amount of matter present in an object. This fundamental physical property identifies how much stuff (substance) there is in a particular physical object. Scientists use the kilogram (kg) as the basic unit of mass in a widely used measurement system called the International System of units *(SI unit system)*. The *pound-mass* (lbm) is the unit of mass in the *American customary system*. It is helpful to remember that 1 kg equals 2.205 lbm.

Since electrically-neutral matter is self-attractive, matter also gives rise to the very important force in nature call gravity (or gravitation). On Earth's surface, people commonly discuss how much mass an object has by expressing its *weight*. This is not scientifically correct because the weight of an object is really a force—defined as the product of the local acceleration of gravity times a mass. (This originates from Newton's second law of motion.) The pioneering work of Galileo Galilei (1564–1642) and Newton in the 17th century led to the classical formulation of gravitation. Early in the 20th century, Einstein provided an expanded interpretation of gravitation within his general relativity theory. Gravitation is the important natural phenomenon that dominates the overall behavior of matter throughout the observable universe.

DENSITY

Ask a person to pick up and describe two blocks of identical volume, one made of aluminum and one made of iron; most likely the person will say that the iron block is "heavier" than the aluminum block. Or else, the person will say that the aluminum block is "lighter" than the iron block. While both responses are correct, neither is particularly useful from a scientific point of view. To assist in more easily identifying and characterizing different materials, scientists devised the macroscopic material property called density. They defined density as the amount of mass contained in a given volume. Modern scientists often use the lower case

Greek letter *rho* (ρ) as the symbol for density in technical publications and equations.

Density is one of the most useful macroscopic physical properties of matter. Solid matter is generally denser than liquid matter; liquid matter

NEWTON'S MECHANICAL UNIVERSE

Newton's laws of motion are the three fundamental postulates that form the basis of the *mechanics* of rigid bodies. He formulated these laws in about 1685, as he was studying the motion of the Moon around Earth and the motions of the planets around the Sun. In 1687, Newton presented this work to the scientific community in *The Principia.*

Newton's first law of motion is concerned with the principle of inertia. It states that if a body in motion is not acted upon by an external force, the momentum remains constant. Scientists also refer to this law as the law of the conservation of momentum.

Newton's second law states that the rate of change of momentum of a body is proportional to the force acting upon the body and is in the direction of the applied force. A familiar statement of this law is the following equation: $F = \mathrm{m}\, a$, where F is the vector sum of the applied forces, m is the mass, and a represents the acceleration vector of the body.

Newton's third law is the principle of action and reaction. It states that for every force acting upon a body, there is a corresponding force of the same magnitude exerted by the body in the opposite direction.

In classical physics, Newton's universal law of gravitation defines gravity as the force of mutual attraction experienced by two masses directed along the line joining their centers of mass. The magnitude of this force is proportional to the product of the two masses and inversely proportional to the square of the distance between the two centers of mass. According to Newton's theory of gravitation, all masses pull on each other with an invisible force commonly called gravity (or gravitation). This force is an inherent property of matter and is directly proportional to an object's mass.

Within Newtonian mechanics, the Sun reaches out across enormous distances in the solar system and "pulls" smaller masses, such as planets, comets, and asteroids, into orbit around it, using its force of gravity. Similarly, Earth "pulls" on its large, natural satellite and keeps the Moon in orbit around the planet. The Earth-Moon orbital relationship is an example of synchronous rotais denser than gases. Scientists use the density of a *material* to determine how massive a given volume of that particular material would be. Density is a function of both the atoms from which a material is composed as well as how closely packed the atoms are arranged in the particular material.

The Earth-Moon system as imaged from space by NASA's *Galileo* spacecraft in 1992 on its way to explore the Jupiter system. Earth's jewel-like beauty contrasts sharply with the Moon's gray, lifeless surface. Earth has a planetary mass of approximately13.16 × 10^{24} lbm (5.97 × 10^{24} kg) and an average density of 344.3 lbm/ft^3 (5.515 g/cm^3). The Moon has a mass of just 16.21 × 10^{22} lbm/ft^3 (7.35 × 10^{22} kg) and an average density of 208.5 lbm/ft^3 (3.34 g/cm^3). It is locked in a synchronous rotation around Earth at an average distance of 238,658 miles (3.84 × 10^5 km). ***(NASA/JPL)***

tion. This means that the Moon has an orbital period (27.322 Earth days) around Earth equal to the amount of time it takes the Moon to rotate on its axis. As a consequence, the Moon always keeps the same side (the near side) toward Earth. Until the space age, the far side of the Moon remained a mystery.

At room temperature (nominally 68°F [20°C]) and one atmosphere pressure, the density of some interesting solid materials found on Earth is as follows: gold 1,205 lbm/ft^3 (19,300 kg/m^3 [19.3 g/cm^3]); iron 493 lbm/ft^3 (7,900 kg/m^3 [7.9 g/cm^3]); diamond (carbon) 219 lbm/ft^3 (3,500 kg/m^3 [3.5 g/cm^3]); aluminum 169 lbm/ft^3 (2,700 kg/m^3 [2.7 g/cm^3]); and bone 112 lbm/ft^3 (1,800 kg/m^3 [1.8 g/cm^3]).

One cubic meter (35.31 ft^3) of any solid or liquid material commonly found on Earth is a large amount of matter—often too much for just one person to handle easily in a laboratory environment. Imagine trying to move a 35.31-ft^3 (1-m^3) chunk of ice (at 32°F [0°C]), which has a mass of about 2,028 lbm (920 kg); or a 35.31-ft^3 (1-m^3) block of pure gold (at 68°F [20°C]), which has a mass of 42,557 lbm (19,300 kg). In research activities, scientists often use smaller quantities of mass and then employ another (equivalent) set of SI units to express the density of a material. Staying within the SI system, they say that ice has a density of 0.92 grams per cubic centimeter (g/cm^3) and gold a density of 19.3 g/cm^3 at the temperature conditions previously mentioned.

The element osmium (Os) is a platinum-family metal with a density of 1,411 lbm/ft^3 (22,600 kg/m^3 [22.6 g/cm^3]) at room temperature. This material has the highest density of any element found naturally on Earth. Osmium is a hard, brittle, lustrous blue-white metal with an *atomic number (Z)* of 76 and an atomic mass (A) of 190. The metal enjoys the highest density title because its atoms are very massive and are packed very closely together in a hexagonal crystalline lattice. For comparison, the metal mercury (Hg), which is a liquid at room temperature, has a density of 849 lbm/ft^3 (13,600 kg/m^3 [13.6 g/cm^3]). At 32°F [0°C] and one atmosphere pressure, liquid water has a density of 62.42 lbm/ft^3 (1,000 kg/m^3 [1.0 g/cm^3]).

Like most gases at room temperature (nominally 68°F [20°C]) and one atmosphere pressure, oxygen (O) has a density of just 0.083 lbm/ft^3 (1.33 kg/m^3 [1.33×10^{-3} g/cm^3])—a value that is about 1,000 times lower than the density of most solid or liquid materials normally encountered on Earth's surface.

It is important to recognize that the physical properties of matter are often interrelated—namely, when one physical property, such as temperature, changes (increases or decreases), other physical properties (such as volume or density) also change. Over the past three centuries, scientists have learned how to define the behavior of materials by developing special mathematical expressions, called equations of state. They developed these mathematical relationships using both theory and empirical data from many carefully conducted laboratory experiments.

Particle physicists use the concept of density in the extreme when they discuss the density of material in an atomic *nucleus.* Assuming an atomic nucleus consists of a collection of *neutrons* and protons, scientists state that the nuclear density of a uranium-235 nucleus is about 1.87×10^{16} lbm/ft^3 (3×10^{17} kg/m^3 [3×10^{14} g/cm^3]).

Astronomers, astrophysicists, and cosmologists also use the concept of density in describing various cosmic objects. Planetary astronomers divide the planets found in humans' solar system into two general classes: terrestrial and jovian. Terrestrial planets, such as Earth, are small, relatively low-mass objects with solid surfaces and high average densities. Jovian planets, such as Jupiter and Saturn, are large, high-total mass, gaseous objects with no solid surfaces and low average densities. For example, the magnificent ringed-world Saturn has an equatorial diameter of 120,536 km, a mass of 12.52×10^{27} lbm (5.68×10^{26} kg), but an average density of just 42.89 lbm/ft^3 (687 kg/m^3 [0.687 g/cm^3]). This means that Saturn has a density less than that of water; so this large, gaseous planet would actually float in some hypothetical giant bowl of water.

A composite image of Saturn that was captured by NASA's *Cassini* spacecraft on February 9, 2004, when the spacecraft was 43.1 million miles (69.4×10^6 km) from the magnificent ringed planet. Saturn has a mass of 12.52×10^{27} lbm (5.68×10^{26} kg) but an average density of just 42.89 lbm/ft^3 (687 kg/m^3 [0.687 g/cm^3]). This means the giant, gaseous planet would hypothetically float in some enormously large, cosmic bowl of water. *(NASA/JPL/Space Science Institute)*

Astrophysicists state that humans' parent star, called Sol by astronomers, has a central (core) density of about 9,364 lbm/ft^3 (150,000 kg/m^3 [150 g/cm^3]), but an average density of just 87.4 lbm/ft^3 (1,400 km/m^3 [1.4 g/cm^3]). At the end of its life, some 5 billion years or so from now, the Sun will run out of hydrogen and ultimately become a "retired star," called a white dwarf. A white dwarf consists of very high density (degenerate) material that has an average estimated density of 6.24×10^8 lbm/ft^3 (1×10^{10} kg/m^3

[1×10^7 g/cm^3]). When a star more massive than the Sun dies in a spectacular Type II supernova event, one of two strange stellar objects is left behind. The first is an extremely high-density remnant known as a neutron star; the other possibility is the creation of a stellar mass *black hole*. With an estimated core density of 6.24×10^{16} lbm/ft^3 (1×10^{18} kg/m^3 [1×10^{15} g/cm^3]), the neutron star is an incredibly dense form of degenerate matter.

The densest imaginable object in the universe is the hypothetical central point within a black hole, called the singularity. Astrophysicists speculate that within a black hole the curvature of space-time is infinite. This

This artist's rendering shows the relative scale of a neutron star to New York City. Scientists do not know if the surface of a neutron star is actually dark gray in color. However, they do know that the extremely dense stellar object has such an intense level gravity that its surface would be the smoothest object in the universe. *(NASA/MSFC)*

causes the singularity to become a point of infinite density and zero volume. White dwarfs, neutron stars, and black holes are discussed in more detail in subsequent chapters. Although such extremely dense objects are beyond day-to-day experience, they play an important role in the functioning of the universe.

At the other end of the density spectrum is the density of interstellar space, estimated to be about 6.24×10^{-22} lbm/ft^3 (1×10^{-20} kg/m^3 [1×10^{-23} g/cm^3]). The best vacuum conditions that can be achieved in laboratories on Earth correspond to a density of about 6.24×10^{-19} lbm/ft^3 (1×10^{-17} kg/m^3 [1×10^{-20} g/cm^3]).

PRESSURE

Scientists define pressure (P) as force per unit area. The most commonly encountered unit of pressure in the American Customary System is *pounds-force* per square inch (psi). In the SI unit system, the fundamental unit of pressure is called the *pascal* (Pa) in honor of the French scientist Blaise Pascal (1623–62), who conducted many pioneering experiments in *fluid mechanics.* One pascal represents a force of one *newton* (N) exerted over an area on one square meter—that is, 1 Pa = 1 N/m^2. One psi is approximately equal to 6,895 Pa.

Anyone who has plunged into the deep end of a large swimming pool and then descended to the bottom of the pool has personally experienced the phenomenon of hydrostatic pressure. Hydrostatic pressure is the pressure at a given depth below the surface of a static (nonmoving) fluid. As Pascal observed in the 17th century, the greater the depth, the greater is the pressure.

On January 23, 1960, the U.S. Navy's deep-diving bathyscaphe *Trieste* successfully submerged and visited the deepest known part of the world's oceans, the Challenger Deep off the Marianas Islands in the Pacific. As part of Project Nekton, the *Trieste* and its two-man crew touched bottom at a depth of 35,838 feet (10,923 m). The hydrostatic pressure at this depth is 16,883 psi (116.4×10^6 Pa). The historic descent into inner space took approximately five hours. Once the *Trieste* arrived at the floor of the Challenger Deep, its crew spent about 20 minutes on the bottom making observations and recording data. Powerful lights enabled the men to observe the seafloor at this incredible depth. The return trip to the ocean's surface took three hours and 17 minutes.

Earth's atmosphere contains mostly nitrogen (78 percent by volume) and oxygen (21 percent by volume). The density of a column of air above a point on the planet's surface is not constant. Rather, density, along with atmospheric pressure, decreases with increasing altitude, until becoming essentially negligible in outer space. The pioneering work of Pascal and the Italian physicist Evangelista Torricelli (1608–47) guided other scientists in measuring and characterizing Earth's atmosphere.

In 1643, Torricelli invented an important instrument called the mercury barometer. At sea level, this instrument indicated the pressure of Earth's atmosphere corresponded to a vertical column of mercury approximately 760 millimeters (mm) in height. To his credit as a scientist, Torricelli observed that the height of the column of mercury in his barometer varied slightly from day to day and correctly concluded that this change in height corresponded to a change in atmospheric pressure. In recognition of his important discovery, scientists defined a special unit of pressure, called the torr. One torr of pressure corresponds to a height of one millimeter of mercury in a barometer or manometer.

Atmospheric pressure plays an important role in many scientific and engineering disciplines. Scientists now use the following values in an effort to standardize their research activities: At sea level on Earth, one atmosphere (1 atm) $\equiv$ 760 mm Hg (exactly) = 29.92 in (Hg) = 14.695 psi = 1.01325×10^5 Pa.

In contrast to Earth's atmosphere, the atmosphere of Venus contains mostly carbon dioxide (CO_2), about 96.5 percent by volume. This enormous amount of CO_2 has caused a runaway greenhouse effect, resulting in an average surface temperature of 873°F (467°C [740 K]). The atmospheric pressure at the surface of this infernolike planet is 1,334 psi (9.2 MPa)—a value that is more than 90 times the sea level atmospheric pressure on Earth.

What are extreme pressures to scientists? Solar physicists estimate that the pressure in the central region (core) of the Sun is approximately 2.90×10^{12} psi (2×10^{16} Pa). Geologists calculate the pressure at the center of planet Earth to be about 5.80×10^7 psi (4×10^{11} Pa). For comparison, a mechanic will typically inflate an automobile tire to about 29 psi (200,000 Pa)—above atmospheric pressure. Finally, the research scientist, who wants no enclosure pressure at all during an experiment on Earth, tries to achieve a "best possible" *vacuum* under laboratory conditions of about 1.45×10^{-16} psi (10^{-12} Pa).

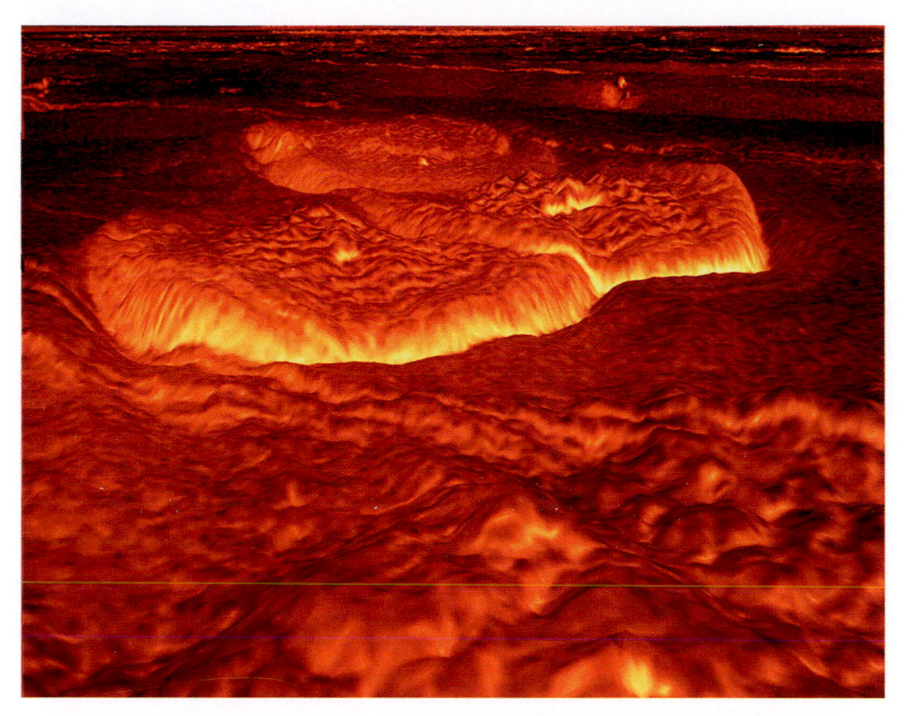

This infernolike landscape is actually a computer-generated, three-dimensional perspective of the Alpha Regio region of Venus. Scientists created this amazing image, which features pancakelike dome volcanoes, by superimposing radar images from NASA's *Magellan* spacecraft and then adding realistic colors, based on surface images from the Russian *Venera 13* and *14* lander spacecraft. When compared to the other terrestrial planets, Venus has the highest atmospheric pressure (more than 90 times that of Earth) and the highest average surface temperature, about 873°F (467°C [740 K])—even hotter than the sunlit surface of Mercury. *(NASA)*

TEMPERATURE

This section discusses the very important macroscopic physical property called temperature (T). While temperature is one of the most familiar physical variables, it is also one of the most difficult to quantify. What is "hot" to one person is just "warm" to another person. Such qualitative interpretations of temperature, though very common, are of little value in science.

Scientists suggest that on a macroscopic scale temperature is the physical quantity that indicates how hot or cold an object is relative to an agreed-upon standard value. Temperature defines the natural direction in which energy will flow as heat—namely from a higher temperature (hot) region to a lower temperature (cold) region. On a microscopic perspective, temperature indicates the speed at which the atoms and molecules of a substance are moving.

The key to quantifying the property of temperature is a physical principle now known as the zeroth law of thermodynamics. This law states that if two objects, called A and B, are each in thermal equilibrium with a third object, called C, then A and B are in thermal equilibrium with each other. In this scenario, the third object (C) is a temperature-measuring device. Thermal equilibrium is a condition that exists when energy transfer as heat between two *thermodynamic systems* is possible but none occurs. The consequences of this simple principle are profound, since temperature is one of the central concepts of *thermodynamics.*

Scientists now recognize that every object has the physical property called temperature. They further state that when two bodies are in thermal equilibrium, their temperatures are equal. A *thermometer* is an instrument that measures temperatures relative to some reference value. As part of the Scientific Revolution, innovative individuals began using a variety of physical principles, natural references, and scales in their attempts to quantify the property of temperature.

In about 1592, Galileo attempted to measure temperature with a device called the thermoscope. Using the principle of buoyancy, he suspended several objects of slightly different densities inside a sealed vertical glass cylinder that contained water. As the room temperature increased, the liquid in the cylinder experienced a rise in temperature and a decrease in density. The less dense objects in the cylinder responded to this change in density by floating toward the top, leaving an observable gap between them and the more dense objects below. The location of this gap (or a single floating object in this gap) provided a rough indication of the temperature of the room. Today, replicas of the Galilean thermometer are available as science demonstration devices and novelty items. In the late 16th century, Galileo's work represented the first serious attempt to harness the notion of temperature as a useful scientific property.

The German physicist Daniel Gabriel Fahrenheit (1686–1736) spent most of his life working in the Netherlands. He was the first person to develop a thermometer capable of making accurate, reproducible measurements of temperature. In 1709, he observed that alcohol expanded when heated, and he constructed the first closed-bulb glass thermometer with alcohol as the temperature-sensitive working fluid. Five years later (in 1714), he used mercury as the thermometer's working fluid. Fahrenheit selected an interesting three-point temperature reference scale for his original thermometers. His zero point (0°F) was the lowest temperature he could achieve with a chilling mixture of ice, water, and ammonium chloride. Fahrenheit then used a mixture of just water and ice as his second reference temperature (32°F). Finally, he chose his own body temperature (recorded as 96°F) as the scale's third reference temperature. After his death, other scientists revised and refined the original Fahrenheit temperature scale, making sure there were 180 degrees between the *freezing point* of water (32°F) and the *boiling point* of water (212°F) at one atmosphere pressure. On this refined scale, the average temperature of the human body now appeared as 98.6°F. Although the Fahrenheit temperature scale is still used in the United States, most of the nations in the world have adopted another temperature scale, called the Celsius scale.

A geologist measures the height of a lava fountain during the 1983 eruption of the Kilauea volcano at the Hawaii Volcanoes National Park. Lava, freshly flowing from a volcanic vent, has a temperature ranging between 1,300°F and 2,200°F (704°C and 1,204°C)—making it the hottest liquid substance found naturally on Earth's surface. *(USGS)*

In 1742, the Swedish astronomer Anders Celsius (1701–44) introduced a temperature scale that is still widely used throughout the world. He initially selected the upper (100°) reference temperature on his new scale as the freezing point of water and the lower (0°) reference temperature as the boiling of water at one atmosphere pressure. He then divided the scale into 100 units. After Celsius's death in 1744, the Swedish botanist and zoologist Carl Linnaeus introduced the modern Celsius scale thermometer by reversing the reference temperatures. The modern Celsius temperature scale is a relative temperature scale in which the range between two reference points (the freezing point of water at 0°C and the boiling point of water at 100°C) are conveniently divided into 100 equal units or degrees.

Despite considerable progress in thermometry in the 18th century, scientists still needed a more comprehensive temperature scale—namely one that included the concept of *absolute zero,* the lowest possible temperature at which molecular motion ceases. In 1848, the Irish-born British physicist William Thomson (1824–1907)—also known as Lord Kelvin, 1st baron of Largs)—proposed an absolute temperature scale. The scientific community quickly embraced Kelvin's new temperature scale.

In the SI unit system, absolute temperature values are expressed in *kelvins* (K), a unit that honors Lord Kelvin. On this scale, absolute zero occurs at 0 K or –273.15°C. By international agreement in 1968, a reference value of 273.16 K, or 0.01°C, was assigned to the triple point of water, in order to create an easily reproducible standard for a one-point absolute temperature scale. In thermodynamics, the triple point is the temperature of a mixture of a substance that contains the liquid, gaseous, and solid

RANKINE—THE OTHER ABSOLUTE TEMPERATURE

Most of the world's scientists and engineers use the Kelvin scale (named after Lord Kelvin, who was also known as William Thomson) to express absolute thermodynamic temperatures. But there is another absolute temperature scale, called the Rankine scale (symbol R), that sometimes appears in engineering analyses performed in the United States—analyses based upon the American Customary System of units.

In 1859, the Scottish engineer and physicist William John Macquorn Rankine (1820–72) introduced the absolute temperature scale that now carries his name. Absolute zero in the Rankine temperature scale (that is, 0 R) corresponds to –459.67 °F. The relationship between temperatures expressed in *rankines* (R) using the (absolute) Rankine scale and those expressed in the degrees Fahrenheit (°F) using the (relative) Fahrenheit scale is: T (R) = T (°F) + 459.67. The relationship between the Kelvin scale and the Rankine scale is: (9/5) × absolute temperature (kelvins) = absolute temperature (rankines). For example, a temperature of 100 K is expressed as 180 R. The use of absolute temperatures is very important in science.

states of the material in thermal equilibrium at standard pressure. The international scientific community uses SI units exclusively, including the absolute temperature scale. The proper SI unit term for temperature is *kelvins* (without the word *degree*), and the proper symbol is K (without the symbol °).

It is interesting to examine some of the temperatures encountered by scientists as they explored the universe. The coldest possible temperature is called absolute zero, a temperature corresponding to 0 rankines (R) (or 0 K). On Earth, scientists have been able to approach the condition of absolute zero in special laboratory environments—reaching temperatures as low as about 1.8×10^{-9} R (10^{-9} K). At the edges of the observable universe, the cosmic microwave background has a measured temperature of about −454°F (–270°C [3 K]). On the surface of Earth, the temperature at which water freezes under normal conditions of atmospheric pressure is 32°F (0°C [273 K]). The temperature of the visible surface of the Sun (a region called the photosphere) is 9,927°F (5,497°C [5,770 K]). Solar physicists estimate that the temperature of the core of the Sun is 27×10^{6} R (15×10^{6} K). Finally, astrophysicists and cosmologists suggest that just after the *big bang* event the universe had a temperature well in excess of 1.8×10^{32} R (1.0×10^{32} K). As discussed in the next chapter, the big bang involved energy and matter under the most extreme conditions ever encountered in the universe.

TIME

The notion of time has always intrigued human beings. Prehistoric peoples used natural phenomena, such as Earth's daily rotation on its axis or the Moon's periodic waxing and waning episodes, to mark its passage. Starting with the ancient Greeks, philosophers have related time with the mind and postulated a form of subjective time. Throughout history, people have felt that pleasant activities were fleeting in their duration, while dull or unpleasant events seemed to drag on forever. However, subjective human experiences in no way influence the actual passage of physical time.

Scientists take a more measured and objective view of physical time. For them, time (most commonly symbolized as t) represents the nonspatial continuum of events, as they proceed from the past through the present and toward the future. The relentless movement of sand in an hourglass vividly depicts this linear notion of time.

Time is an important physical quantity in science. The hourglass appeared in Europe in the 1300s and was used to monitor short intervals of time, typically an hour or so. The device now serves as a universal symbol for the passage of time. The sand at the top represents the future; the sand falling through the neck is the present; and the sand that has accumulated at the bottom is the past. *(DOE/FNAL)*

Time flows from the past, through the present, to the future. Scientists use time to quantify the motion of matter in the universe. They also use time to describe the interval between events or physical activities, the sequence of phenomena in a certain process, and the overall interval needed for a certain event to occur.

Galileo and Newton were the first scientists to formally include time in the study of the motion of matter. Although Galileo pioneered the investigation of the periodic behavior of a pendulum, his research in mechanics remained severely hampered by his inability to accurately record the passage of time. Newton developed a new form of mathematics, called calculus, to handle time-dependent problems involving changes in an object's position, velocity, or *acceleration*. The German mathematician Gottfried Leibniz (1646–1716) also receives credit for independently developing calculus. In the 1650s, the Dutch scientist Christiaan Huygens (1629–95) constructed and patented the first pendulum clock. His efforts greatly improved scientific timekeeping

Today, scientists typically treat time as the duration between two instants. The SI unit of time is the second (s), which now is precisely defined using an atomic standard. Time also may be designated as solar time, lunar time, or sidereal time, depending on the astronomical reference used. For example, solar time is measured by successive intervals between transits of the Sun across the meridian. Lunar time involves phases of Earth's Moon. The lunar month, for example, represents the time taken by the Moon to complete one revolution around Earth, as measured from new Moon to new Moon. The lunar month is also called the synodic month and is 29.5306 days long. Sidereal time is measured by successive transits of the vernal equinox across the local meridian. On

Earth, time can be local or associated with the worldwide scientific standard of timekeeping, called universal time coordinated (UTC). Universal time coordinated is based on carefully maintained atomic clocks and uses Greenwich, England, as a reference point.

Several time intervals are of special interest to scientists. Named after the German physicist Max Planck (1858–1947), who founded quantum theory, the Planck time (1.3×10^{-43} s) is the incredibly short interval of time associated with the Planck era, the first cosmic epoch, when (as physicists now hypothesize) quantum gravity controlled the very young universe immediately following the big bang. Scientific theory is currently inadequate to describe the conditions of the very early universe, when the temperature was above 1.8×10^{32} R (1×10^{32} K) and all nature's four forces were believed to be merged into one single superforce.

The age of the universe is 13.7 billion years; the solar system is about 5 billion years old. Solar physicists currently estimate that the Sun will continue its dependable thermonuclear conversion of hydrogen into helium for another 5 billion years. Some physicists support theories of matter that require protons to ultimately fall apart (that is, experience decay); such scientists are now trying to obtain credible experimental evidence to support their hypothesis. The theoretically postulated *half-life* of the proton is on the order of 10^{32} years. If protons really are unstable and experience decay, then by the time the universe reaches the age of about 10^{40} years, all atomic matter (such as contained in stars and planets and giant molecular clouds of interstellar gas), will have disintegrated into *subatomic particles* and radiation.

When an object experiences uniform acceleration (a), scientists define its velocity (v) as the rate of uniform acceleration (a) multiplied by time (t) the object experienced that acceleration. Scientists use mathematics to express this simple, yet very important rate equation as: $v = a \times t$. When an object travels at constant velocity, they use another basic rate equation to calculate the total range or distance traveled (d) during a period of time (t). The range equation is simply: $d = v \times t$. Starting with Newton and Leibniz, scientists also began using calculus to deal with more complicated matter-motion relationships, including those involving time-dependent velocities, accelerations, and masses. As part of the Scientific Revolution, 17th-century scientists quantified time and included this elusive, nonspatial quantity along with the three spatial dimensions—commonly represented as x-, y-, and z- in the Cartesian coordinate system. In classical physics, scientists treated time intervals as absolute quantities.

THE PRIMARY TIME AND FREQUENCY STANDARD FOR THE UNITED STATES

Called NIST-F1, this cesium fountain atomic clock is the nation's primary time and frequency standard. The extremely precise timekeeping device was developed by the laboratories of the National Institute of Standards and Technology (NIST) in Boulder, Colorado, based on the microwave transitions in the cesium-133 atom. NIST-F1 contributes to the internal group of atomic clocks that define UTC, the official world time. Because NIST-F1 is among the most accurate clocks in the world, the NIST device makes UTC more accurate than ever before. For example, in 2005, the NIST-F1 atomic clock would neither gain nor lose a second in more than 60 million years; the device is continually being improved by NIST scientists.

Scientists refer to NIST-F1 as a fountain clock because the device uses a fountainlike movement of atoms to measure frequency and time interval. First, a collection of cesium atoms (in gaseous form) is introduced into the clock's vacuum chamber. Six infrared laser beams are then directed at right angles to one another at the center of the chamber. The infrared lasers gently push the cesium atoms together into a ball. In the process of creating this ball, the lasers slow down the movement of the atoms and cool them to temperatures near absolute zero. With all the other lasers turned off, the two vertical lasers gently toss the extremely cold ball of cesium atoms upward, producing a "fountain" action. This little nudge is just enough to loft the ball of atoms about three feet (1 m) high through a microwave-filled cavity. Under the influence of gravity the ball then falls back down through the microwave cavity.

It takes about one second for the ball of cesium atoms to make the trip up and down through the microwave cavity. During the trip, the atomic states

In 1905, Einstein upset the world of classical physics by introducing the notion of time dilation with his special theory of relativity. For Einstein, the intervals of time were not absolute but relative to the motion of the observers. Special relativity is discussed in subsequent chapters. For now, it is sufficient to note that two identical atomic clocks, one on the surface of Earth and one a satellite in orbit around Earth, will record different time intervals. The clock traveling on the satellite around Earth, will actually record less elapsed time (that is, run slower) than the stationary clock of the cesium atoms might or might not be altered as they interact with the microwave signal. When their trip is completed, another laser is pointed at the cesium atoms. Any atom whose atomic state was altered by the microwave signal now emits light (an excited state of matter called fluorescence). Detectors then measure the emitted photons.

This process is repeated many times while scientists tune the microwave signal in the cavity to different frequencies. Scientists eventually find the microwave frequency that alters the states of most of the cesium atoms in the ball and maximizes their fluorescence. This frequency corresponds to the natural resonance frequency of the cesium atom (9,192,631,770 *hertz*), or the frequency used to define the second. Improved frequency control results in one of the world's most accurate atomic clocks.

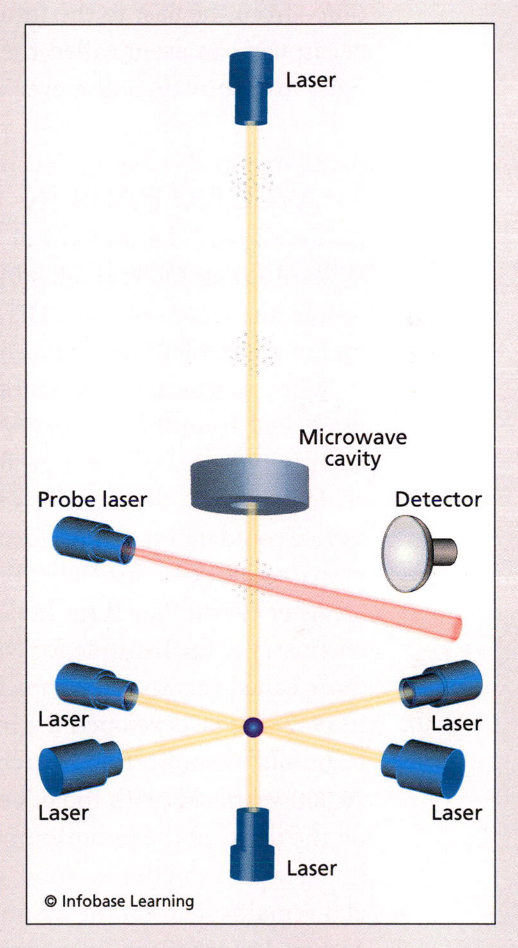

This illustration depicts the fundamental components of the NIST-F1 Cesium Fountain Clock—the primary time and frequency standard for the United States. *(NIST)*

on Earth. The fascinating time dilation aspects of special relativity have been demonstrated by numerous experiments, including experiments involving the use of identical atomic clocks—one kept stationary on Earth and the other placed onboard an orbiting spacecraft.

Both scientists and philosophers puzzle over questions such as why people remember the past and not the future, or why a person can transform a fresh egg into an omelet, but an omelet never transforms itself into a whole egg (including shell). Irreversible processes, *entropy,* and the second law of thermodynamics lie at the heart of an interesting concept called the arrow of time. Observations of phenomena in the physical universe indicate that time behaves like an arrow pointing in only one direction—from the past to the future. As discussed in the next chapter, time began with an event called the big bang and has pointed in the forward (past-to-future) direction ever since.

TRAVELING BACK IN TIME USING RADIOACTIVITY

By developing the method of carbon-14 dating, the American chemist Willard Frank Libby (1908–80) used the phenomenon of *radioactivity* to unlock the secrets of time. His discovery provided an important research tool in archaeology, geology, and other branches of science.

Libby was born on December 17, 1908, in Grand Valley, Colorado, and graduated from the University of California at Berkeley with a Ph.D. in chemistry in 1933. While performing his doctoral research, Libby constructed one of the first Geiger-Müller tubes in the United States. The device could detect certain forms of *nuclear radiation,* and the experience prepared Libby for his Nobel Prize–winning research in the late 1940s.

After World War II (in 1945), Libby accepted a position as professor of chemistry at the Institute for Nuclear Studies of the University of Chicago (now called the Enrico Fermi Institute for Nuclear Studies). It was there in 1947 that he developed the innovative concept for the carbon-14 dating technique—a powerful research tool for reliably dating objects up to about 70,000 years old (with today's equipment). He based the clever technique on the decay of the radioisotope carbon 14, as contained in such formerly living organic matter as wood, charcoal, parchment, shells, and even skeletal remains; and on the assumption that the absorption of atmospheric carbon 14 (produced by cosmic-ray interactions in Earth's atmosphere) ceases when a living thing dies. At that point, the radioactive decay clock in the organic matter starts ticking. Scientists can determine an object's age by comparing its (reduced) carbon-14 activity levels to those (higher) carbon-14 activity levels found in comparable living organisms or viable organic materials.

In developing his idea, Libby reasoned that the level of carbon-14 radioactivity in any sample of organic material should clearly indicate the time of the organism's death. The real challenge he faced and overcame was to develop and operate a sensitive enough radiation detection instrument that was capable of accurately counting the relatively weak beta decay events of carbon 14. With a half-life of 5,730 years, radiocarbon is only mildly radioactive.

Libby constructed a sufficiently sensitive Geiger counter. He then successfully tested his proposed carbon-dating technique against organic objects from antiquity—objects that had reasonably well-known ages. For example, he successfully used radiocarbon dating to determine the ages of a wooden boat from the tomb of an Egyptian pharaoh, a piece of prehistoric sloth dung that was found in Chile, and a wrapping from the Dead Sea Scrolls. His research demonstrated that carbon-14 analysis represented a reliable way of dating organic objects from the past. Libby received the 1960 Nobel Prize in chemistry for developing the method of carbon-14 dating and for its many important applications in archaeology, geology, and other branches of science. He summarized this interesting use of carbon in the 1952 book *Radiocarbon Dating.*

CHAPTER 2

Birth of the Universe

About 13.7 billion years ago, an incredibly powerful explosion took place. Called the big bang, this enormous explosion created space and time, as well as all the energy and matter that now exists in the universe. This chapter summarizes the latest scientific understanding about this extreme event. Physicists are still not sure what "banged" or why it "banged." But immediately after the big bang, the universe began to expand, and energy and matter began a fascinating series of transformations that have led to the universe in its present state, which includes stars, planets, and intelligent life. This chapter discusses the origin of ordinary (or baryonic) matter within the context of big bang cosmology. A brief mention is also made of the phenomenon of *dark energy* and how it, along with dark matter, has helped define the currently observed universe.

BIG BANG COSMOLOGY

The big bang is a widely accepted theory in contemporary cosmology, concerning the origin and evolution of the universe. According to the big bang cosmologists, about 13.7 billion years ago there was an incredibly powerful explosion that started the present universe. Before this ancient explosion, matter, energy, space, and time did not exist. All of these physical phenomena emerged from an unimaginably small, infinitely dense object, which physicists call the initial singularity. Immediately after the

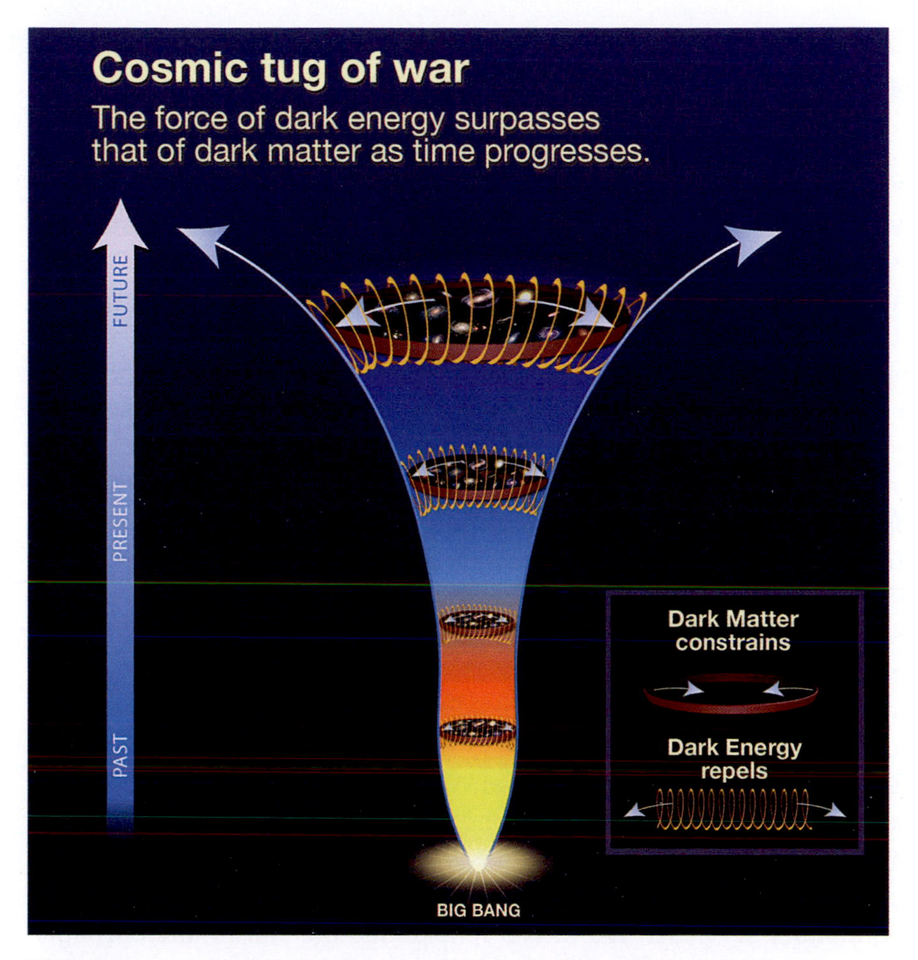

This diagram suggests that the overall history of the universe can be envisioned as a cosmic tug-of-war between the gravitational pull of dark matter (the universe's major form of matter) and the push of dark energy (a mysterious phenomenon of nature). This process began at least 9 billion years ago, well before dark energy gained the upper hand and began accelerating the expansion of the universe. *(NASA, ESA, and A. Feild [STScI])*

big bang, the intensely hot, infant universe began to expand and cool. As the temperature dropped, energy and matter started to experience several fascinating transformations.

Astrophysical observations indicate that the universe has been expanding ever since the big bang but at varying rates of acceleration. The continuous cosmic tug-of-war between the mysterious pushing force of dark

(continues on page 28)

HUBBLE AND THE EXPANDING UNIVERSE

In the mid-1920s, the American astronomer Edwin Powell Hubble dramatically revised the scientific understanding of the universe by proving that galaxies existed beyond the Milky Way galaxy. This revolution in cosmology started in 1923, when he used the periodic pulsation pattern of a Cepheid variable star to estimate the distance to the Andromeda galaxy. His results immediately suggested that such spiral nebulas were actually large, distant independent stellar systems, or island universes. Next, he introduced a classification scheme in 1925 for such "nebulas" (galaxies), calling them either elliptical, spiral, or irregular. Finally, in 1929, Hubble announced that the universe was expanding. In this expanding universe, the other galaxies were receding from Earth with speeds proportional to their distance—a postulate now known as Hubble's Law. Hubble's concept of an expanding universe filled with numerous galaxies forms the basis of modern observational cosmology.

Hubble was born in Marshfield, Missouri, on November 20, 1889. Early in his life, Hubble studied law at the University of Chicago and then at Oxford. In 1913, he abandoned law and pursued astronomy. From 1914 to 1917, Hubble held a research position at the Yerkes Observatory, operated by the University of Chicago. There on the shores of Lake Geneva at Williams Bay, Wisconsin, he began investigating interesting nebulas. By 1917, he concluded that the spiral-shaped ones (astronomers now call them galaxies) were quite different from the diffuse nebulas, which are actually giant clouds of dust and gas.

Following military service in World War I, he joined the staff at the Carnegie Institute's Mount Wilson Observatory, located in the San Gabriel Mountains, northwest of Los Angeles. When Hubble arrived in 1919, this observatory's 100-inch (2.54-m) diameter telescope was the world's biggest optical astronomy instrument. Except for other scientific work during World War II, Hubble remained affiliated with this observatory for the remainder of his life. Once at Mount Wilson, Hubble used its large instrument to resume his careful investigation of nebulas.

In 1923, Hubble discovered a Cepheid variable star in the Andromeda nebula—a celestial object now known to astronomers as the Andromeda galaxy, or M31. A Cepheid variable is one of a group of important very bright supergiant stars that pulsate periodically in brightness. By carefully studying this particular Cepheid variable in M31, Hubble was able to conclude that it was very far away and belonged to a separate collection of stars far beyond the Milky Way

This is a composite image of the Pinwheel galaxy, also called M101 or NGC 5457. Located about 25 million light-years from Earth in the constellation Ursa Major, this galaxy is approximately 170,000 light-years across. NASA scientists created the composite image with data from the *Hubble Space Telescope*, the *Spitzer Space Telescope*, and the *Chandra X-ray Observatory*. The red color depicts *Spitzer*'s view in infrared light; the yellow color shows *Hubble*'s view in visible light; and the blue color shows *Chandra*'s view in X-ray light. *(NASA/JPL-Caltech/ESA/CXC/STSci)*

galaxy. This important discovery provided the first tangible observational evidence that galaxies existed beyond the Milky Way. Through Hubble's pioneering efforts, the human perception of the size of the known universe expanded by incredible proportions.

Hubble continued to study other galaxies and in 1925 introduced the following well-known classification system: spiral galaxies, barred spiral galaxies, elliptical galaxies, and irregular galaxies. Then, in 1929, Hubble investigated the

(continues)

(continued)

recession velocities of galaxies (that is, the rate at which galaxies are moving apart) and their distances away. He discovered that the more distant galaxies are receding (going away) faster than the galaxies closer to Earth. This very important discovery revealed that the universe is expanding. Hubble's work provided the first observational evidence that supported big bang cosmology. The concept of the universe expanding at a uniform, steady rate was codified in a simple mathematical relationship called Hubble's Law in his honor.

Hubble's Law describes the expansion of the universe as a linear, steady rate. As Hubble initially observed and subsequent astronomical studies confirmed, the apparent recession velocity (v) of galaxies is proportional to their distance (r) from an observer. The proportionality constant is H_0, the Hubble constant. Currently, proposed values for H_0 fall between 50 and 90 kilometers per second per megaparsec (km s^{-1} Mpc^{-1}). The inverse of the Hubble constant ($1/H_0$) is called the Hubble time. Astronomers have used the value of the Hubble time as one measure of the age of the universe.

Cosmologists and astrophysicists employed several techniques to discern that the universe is about 13.7 billion years old. Precise astrophysical observations of distant Type Ia supernovas made in the late 1990s indicate that rate of expansion of the universe is actually increasing. These recent data affect the linear expansion rate hypothesis inherent in Hubble's Law. However, the recent observations, while very significant, take very little away from Hubble's brilliant pioneering work. Hubble provided the first observational evidence of an enormously large, expanding universe filled with billions of galaxies.

Quite fittingly, NASA named the *Hubble Space Telescope* after him. This powerful orbiting astronomical observatory continued the fine tradition of extragalactic investigation started by Hubble, who died on September 28, 1953, in San Marino, California.

(continued from page 25)

energy (discussed shortly) and the gravitational pulling force of matter (both ordinary matter and dark matter) appears responsible for this variation in rate of acceleration.

In 1927, the Belgian astrophysicist, cosmologist, and Roman Catholic priest Georges-Édouard Lemaître (1894–1966) became the first scientist to formally suggest that a violent explosion might have started an expanding universe. He based this interesting hypothesis on his own scientific interpretation of Einstein's general relativity theory and its cosmological implications. Central to Lemaître's theoretical model was the idea of an initial cosmic egg or superdense primeval atom that started the universe in a colossal ancient explosion. Other scientists, such as the Ukrainian-American physicist George Gamow (1904–68), built upon Lemaître's work and developed the cosmic egg concept into the now widely accepted big bang theory of modern cosmology.

Two years later (in 1929), the American astronomer Edwin P. Hubble (1889–1953) announced that the universe was expanding. He based this bold statement upon his personal observations of Doppler-shifted *wavelengths* of the light from distant galaxies, which indicated that these galaxies were receding from Earth with speeds proportional to their distance.

The key observation that supported big bang cosmology took place in 1964, when the German-American physicist Arno Allen Penzias (1933–) and the American physicist Robert Woodrow Wilson (1936–) detected the cosmic microwave background (CMB). Physicists regard the CMB as the remnant radiation signature of an intensely hot, young universe. Although the discovery occurred quite by accident, their pioneering work provided the first direct scientific evidence that the universe had an explosive beginning.

Prior to the discovery of the CMB, the majority of scientists were philosophically comfortable with a *steady-state* model of the universe—a model postulating that the universe had no beginning and no end. Within this steady-state model, scientists assumed matter was being created continuously (by some undefined mechanism) to accommodate the universe's observed expansion.

In 1949, the British astronomer Sir Fred Hoyle (1915–2001), who helped develop and strongly advocated the steady-state model, coined the term *big bang.* Hoyle intended the term as a derisive expression against fellow astronomers, who favored big bang cosmology. Unfortunately, Hoyle's derogatory use of the term backfired. The expression immediately gained favor among those competing scientists, who warmly embraced the term *big bang* as a clever way for them to succinctly explain the new theory. Detailed observations of the cosmic microwave background (CMB) proved that the universe did indeed have an explosive beginning about 13.7 billion years ago. As evidence mounted in favor of big bang cosmology, the vast majority of scientists abandoned the steady-state universe model.

Near the end of the 20th century, big bang cosmologists had barely grown comfortable with a gravity-dominated, expanding universe model when a major cosmological surprise appeared. In 1998, two competing

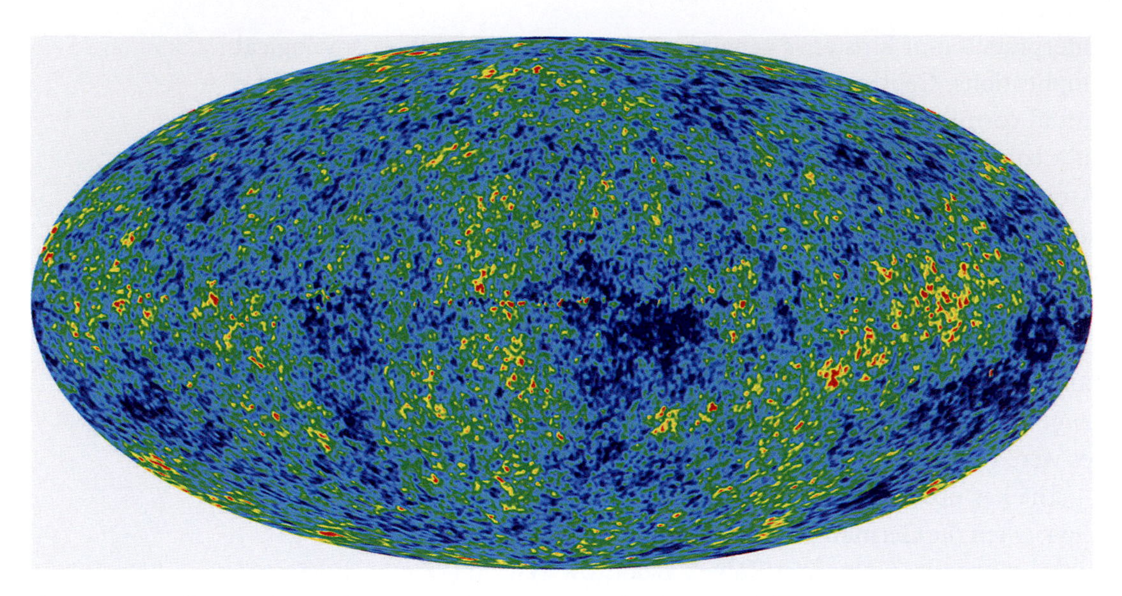

Temperature fluctuations of the cosmic microwave background. The average temperature is 2.725 kelvins (K), and the colors represent the tiny temperature fluctuations. Red regions are warmer and blue regions are colder by about 0.0002 K. This full sky map of the heavens is based on five years of data collection by NASA's *Wilkinson Microwave Anisotropy Probe (WMAP). (NASA/WMAP Science Team)*

teams of astrophysicists independently observed that the universe was not just expanding, but expanding at an increasing rate.

Dark energy is the generic name that astrophysicists have given to the unknown cosmic force field believed responsible for the acceleration in the rate of expansion of the universe. In the late 1990s, scientists performed systematic surveys of very distant Type Ia (carbon detonation) supernovas. They observed that instead of slowing down (as might be anticipated if gravity was the only significant force at work in cosmological dynamics), the rate of recession (that is, the redshift) of these very distant objects appeared to be actually increasing. It was almost as if some unknown force was neutralizing or canceling the attraction of gravity. Such startling observations proved controversial and very inconsistent with the then standard gravity-only models of an expanding universe within big bang cosmology. Today, carefully analyzed and reviewed Type 1a supernova data indicate that the universe is definitely expanding at an accelerating rate.

Physicists do not yet have an acceptable answer as to what phenomenon is causing this accelerated rate of expansion. Some scientists have

HISTORY OF THE UNIVERSE

The accompanying illustration depicts the evolution of the universe over the past 13.7 billion years from the big bang event to the present day—symbolized by NASA's *Wilkinson Microwave Anisotropy Probe (WMAP)* spacecraft. The far left side of this illustration shows the earliest moment scientists can now investigate—the time when an unusual period of inflation produced a burst of exponential growth in the young universe. (NASA illustrators have portrayed the size of the universe by the vertical extent of the grid and age of the universe in the horizontal direction, starting with the big bang and the beginning of time.) For the next several billion years, the expansion of the universe gradually slowed down as the matter in the universe (both ordinary and dark) tugged on

(continues)

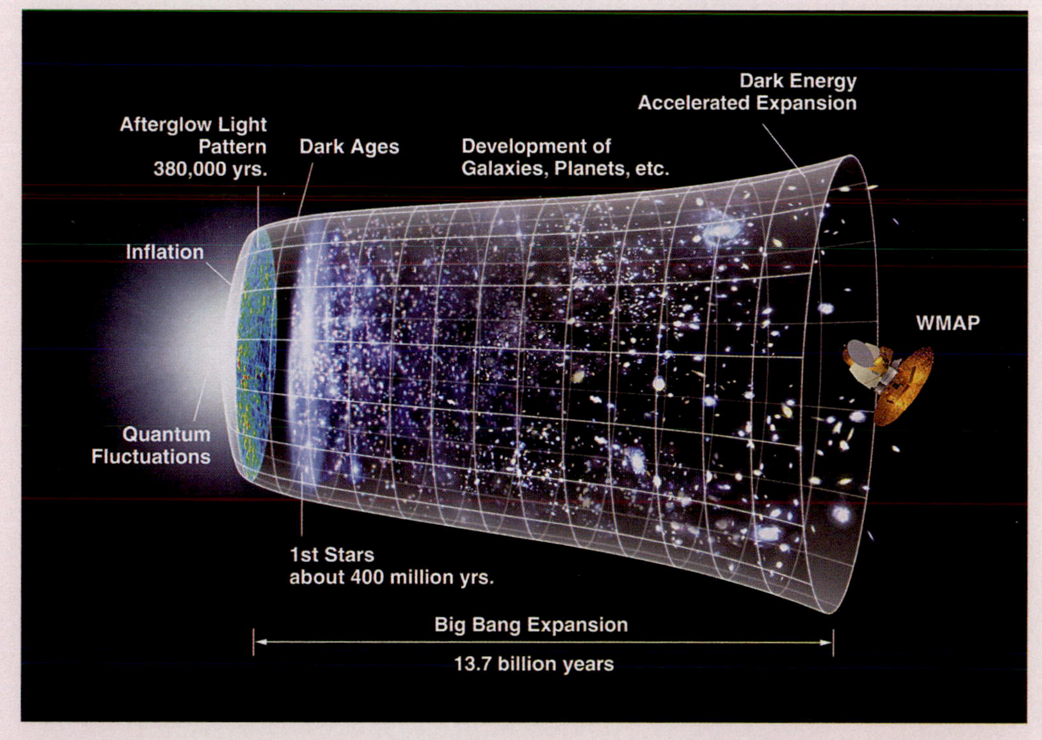

This figure provides a time line of the universe. Events are based on scientific data, including detailed measurements of the cosmic microwave background as performed by NASA's *Wilkinson Microwave Anisotropy Probe (WMAP)* spacecraft. *(NASA/WMAP Science Team)*

(continued)

itself through gravitational attraction. More recently, the expansion of the universe started to speed up again, as the repulsive effects of dark energy became dominant.

The afterglow light seen by NASA's *WMAP* spacecraft was emitted about 380,000 years after the rapid expansion event called inflation and has traversed the universe largely unimpeded since then. When scientists examined the CMB carefully, they discovered telltale signatures imprinted on the afterglow light, revealing information about the conditions of the very early universe. The imprinted information is helping scientists understand how primordial events caused later developments (such as the clustering of galaxies), as the universe expanded.

When scientists peer deep into space with their most sensitive instruments, they also look far back in time—eventually viewing the moment when the early universe transitioned from an opaque gas to a transparent gas. Beyond that transition point, any earlier view of the universe remains obscured. The CMB serves as the very distant wall that surrounds and delimits the edges of the observable universe. Scientists can only observe cosmic phenomena back to the remnant glow of this primordial hot gas from the big bang. The afterglow light experienced Doppler shift to longer wavelengths due to the expansion of the universe. The CMB currently resembles emission from a cool dense gas at a temperature of only 4.905 R (2.725 K).

The very subtle variations observed in the CMB are challenging big bang cosmologists. They must explain how clumpy structures of galaxies could have evolved from a previously assumed smooth (that is, uniform and homogeneous) big bang event. Launched on June 30, 2001, NASA's *WMAP* spacecraft made a very detailed full-sky map of the cosmic microwave background—including high-resolution data of those subtle CMB fluctuations, which represent the primordial seeds that germinated into the cosmic structure scientists observe today. The patterns detected by *WMAP* represent tiny temperature differences within the very evenly dispersed microwave light bathing the universe—a light that now averages a frigid 4.905 R (2.725 K). The subtle CMB fluctuations have encouraged physicists to make modifications in the original big bang hypothesis. One of these modifications involves a concept they call inflation. During inflation, vacuum state fluctuations gave rise to the very rapid (exponential), nonuniform expansion of the early universe.

revisited the cosmological constant (symbol Λ), which Einstein inserted into his original general relativity theory to make that theory of gravity describe a static universe. A static universe would be a nonexpanding one that had neither a beginning nor an end. After boldly introducing the cosmological constant as representative of some mysterious force associated with empty space capable of balancing or even resisting gravity, Einstein abandoned the idea. Lemaître's theoretical astrophysical work and Hubble's announcement of an expanding universe provided the intellectual nudge that encouraged his decision. Later in life, Einstein would refer to his postulation of a cosmological constant as "my greatest failure."

But Einstein may have been on the right track all along. Physicists are now revisiting Einstein's concept and suggesting that there is possibly a vacuum pressure force (a recent name for the cosmological constant), which is inherently related to empty space but exerts its influence only on a very large scale. The influence of this mysterious force would have been negligible during the very early stages of the universe following the big bang event but would later manifest itself and serve as a major factor in cosmological dynamics. Since such a mysterious force is neither required nor explained by any of the currently known laws of physics, scientists do not yet have a clear physical interpretation of what this gravity-resisting force means. Dark energy is discussed further in chapter 6.

THE FIRST THREE MINUTES

Current levels of science are inadequate to describe the extreme conditions of the very young universe during the Planck era—that is, during the first 10^{-43} s after the big bang. Scientists speculate that in the Planck era the very young universe was extremely small and extremely hot, perhaps at temperatures greater than 1.8×10^{32} R (1.0×10^{32} K). At such unimaginably high temperatures, the four fundamental forces of nature were most likely merged into one, unified superforce.

Limitations imposed by the uncertainty principle of quantum mechanics currently prevent scientists from developing a model of the universe between the big bang event and the end of Planck time. This dilemma has forged an interesting intellectual alliance between astrophysicists (who study the behavior of matter and energy in the large-scale universe) and high-energy nuclear particle physicists (who investigate the behavior of matter and energy in the subatomic world).

Particle physicists currently postulate that the early universe experienced a specific sequence of phenomena and phases following the big bang explosion. They generally refer to this sequence as the standard cosmological model. Right after the big bang, the universe was at an incredibly high temperature, perhaps above the unimaginable value of 1.8×10^{32} R (1.0×10^{32} K). During this period, sometimes called the quantum gravity epoch, the force of gravity, the strong force, and a composite electroweak force all behaved as a single unified force. Scientists postulate that at this time the physics of elementary particles and the physics of space and time were one and the same. In an effort to adequately describe quantum gravity, some physicists are trying to develop a theory of everything (TOE).

About 10^{-43} second after the big bang, the force of gravity assumed its own identity. With a temperature estimated to be about 1.8×10^{32} R ($1.0 \times$

FOUR FUNDAMENTAL FORCES IN NATURE

At present, physicists recognize the existence of four fundamental forces in nature: gravity, electromagnetism, the *strong force,* and the *weak force.* Gravity and electromagnetism are part of a person's daily experiences. These two forces have an infinite range, which means they exert their influence over great distances. For example, the two stars in a binary star system experience mutual gravitational attraction even though they may be a *light-year* or more distant from each other. Gravity is the only known force that scientists treat as significant over astronomical (cosmic) distances. Gravity allows mass to shape the observable universe.

The other two forces, the strong force and the weak force, operate within the realm of the atomic nucleus and involve *elementary particles.* These forces lie beyond daily human experiences. The strong and weak forces remained unknown to physicists of the 19th century, despite their good classical understanding of the law of gravity and the fundamental principles of electromagnetism. The strong force operates at a range of about 3.28×10^{-15} ft (1.0×10^{-15} m) and holds the atomic nucleus together. The weak force has a range of about 3.28×10^{-17} ft (1.0×10^{-17} m) and is responsible for processes such as beta decay that tear nuclei and elementary particles apart. What is important to recognize here is that whenever anything happens in the universe—that is, whenever an object experiences a change in motion—the event takes place because one or more of these fundamental forces is involved.

10^{32} K), the entire spatial extent of the universe at that moment was less than the size of a proton. Scientists call the ensuing period the grand unified epoch. While the force of gravity functioned independently during the grand unified epoch, the strong force and the composite electroweak force remained together and continued to act like a single force. Physicists apply various grand unified theories (GUTs) in their efforts to model and explain how the strong force and the electroweak force functioned as one during this particular period in the early universe.

At 10^{-35} second after the big bang, the strong force separated from the electroweak force. By this time, the expanding universe had "cooled" to about 1.8×10^{28} R (1.0×10^{28} K). Two other important activities took place around this time: the annihilation of antimatter and inflation (the incredibly rapid expansion of space). Scientists think that in the very early universe about 10,000,001 protons formed for every 10,000,000 antiprotons that formed. Since a proton and an antiproton annihilate each other on contact, the early universe soon became flooded with an intense amount of annihilation radiation. The surviving protons (and other ordinary particles such as electrons and neutrons) subsequently formed all the matter in the observable universe (see chapter 8).

Physicists call the period between about 10^{-35} s and 10^{-10} s the electroweak force epoch. During this epoch, the weak force and the electromagnetic force became separate entities, as the composite electroweak force disappeared. From this time forward, the universe contained the four fundamental forces in nature now known to physicists—namely, the force of gravity, the strong force, the weak force, and the electromagnetic force.

Following the big bang and lasting up to about 10^{-35} s, there was no distinction between *quarks* and *leptons*. All the minute particles of matter were similar. Then, during the electroweak epoch, quarks and leptons became distinguishable. This transition allowed quarks and antiquarks to eventually become *hadrons,* such as neutrons and protons, as well as their antiparticles. At 10^{-4} s after the big bang, in the radiation-dominated era, the temperature of the universe cooled to 1.8×10^{12} R (1.0×10^{12} K). By this time, most of the hadrons disappeared because of matter-antimatter annihilations. The surviving protons and neutrons represented only a small fraction of the total number of particles in the universe—the majority of which was leptons, such as electrons, *positrons,* and *neutrinos.* Like most of the hadrons before them, most of the leptons soon disappeared as a result of matter-antimatter interactions.

INFLATION

In cosmology, inflation is the theorized exponential expansion of the very early universe that took place about 10^{-35} s after the big bang. The American physicist Alan Harvey Guth (1947–) was one of the first scientists to propose the concept of inflation. He did this in 1980 in order to help solve some of the lingering problems in cosmology that were not adequately resolved in standard big bang cosmology. One of the perplexing questions involved the origin of the subtle density enhancements that resulted in the formation of the galaxies and larger structures (such as clusters of galaxies) now observed in the universe. Another pressing question involved the large-scale uniformity of the universe. For example, cosmic microwave background data suggest that at the end of the era of nuclei (about 380,000 years ago), the density of the universe varied from place to place by no more than approximately 0.01 percent. The final question that challenged cosmologists was the one involving the density of the universe and why it was so close to the critical density.

As Guth and other scientists suggested, between 10^{-35} and 10^{-33} seconds after the big bang, the universe actually expanded at an incredible rate—far in excess of the speed of light. In this very brief period, the universe increased in size by at least a factor of 10^{30}—from an infinitesimally small subnuclear-size dot to a region about nine feet (3 m) across. By analogy, imagine a grain of very fine sand becoming the size of the presently observable universe in one-billionth (10^{-9}) the time it takes light to cross the nucleus of an atom. During

At the beginning of the matter-dominated era, some three minutes (or 180 seconds) following the big bang, the expanding universe cooled to a temperature of 1.8×10^9 R (1.0×10^9 K) and low-mass nuclei, such as deuterium, helium-3, helium-4, and a trace quantity of lithium-7 began to form as a result of very energetic collisions in which the reacting particles joined or fused. Once the temperature of the expanding universe dropped below 1.8×10^9 R (1.0×10^9 K), environmental conditions were no longer favorable to support the fusion of other low-mass nuclei and the process of big bang nucleosynthesis ceased. The observed amount of helium in the universe agrees with the theoretical predictions of big bang nucleosynthesis.

Later, when the expanding universe reached an age of about 380,000 years, the temperature had dropped to about 5,400 R (3,000 K), allowing inflation, space itself expanded so rapidly that the distances between points in space increased greater than the speed of light. Scientists speculate that the slight irregularities they now observe in the cosmic microwave background are evidence (the fossil remnants or faint ghosts) of the ancient quantum fluctuations that took place as the early universe inflated.

Inflation theory resolved many of the problems scientists encountered with standard big bang cosmology. For example, inflation explained the universe's large-scale uniformity by suggesting that currently observed distant regions were once close enough (prior to inflation) to have exchanged radiation, thereby equalizing their temperatures and densities. Specifically, scientists postulate that when the universe was less than 10^{-38} s old, different regions were typically less than 10^{-38} light-second apart. So, radiation traveling at the speed of light would have had sufficient time to travel back and forth between different spatial regions, equalizing temperature and density.

Cosmologists define critical density as the precise average value of density for the entire universe (including ordinary matter, dark matter, and dark energy), which serves as the dividing line between a universe that will expand forever (an open universe) and one that collapses back into a singularity (a closed universe). Scientists supporting inflation theory point out that the overall density of matter and energy in the universe corresponds to the critical density. This implies that the overall geometry of the universe is critical or flat. (Readers desiring a more detailed discussion of inflation are referred to any recent college-level astronomy textbook.)

electrons and nuclei to combine and form hydrogen and helium atoms. Interstellar space remains filled with the remnants of these primordial hydrogen and helium atoms. Within the period of atom formation, the ancient fireball became transparent and continued to cool from a hot 5,400 R (3,000 K) down to a frigid 4.905 R (2.725 K)—the currently measured value of the cosmic microwave background (CMB).

In an effort to understand conditions of the very early universe, researchers are trying to link the physics of the very small (as described by *quantum mechanics*) with the physics of the very large (as described by Einstein's general relativity theory). What scientists hope to develop is a new realm of physics called quantum gravity, which could be capable of treating Planck-era phenomena. (Chapter 3 discusses the world of the very small)

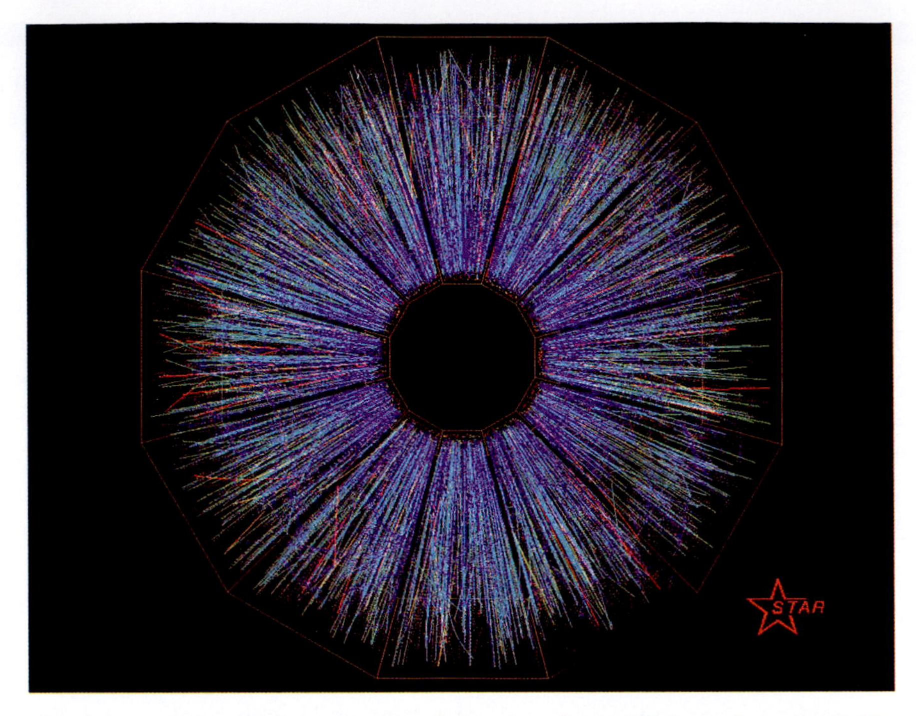

This image shows the results of a violent head-on collision of two 30 GeV beams of gold atom nuclei in the Relativistic Heavy Ion Collider (RHIC) at Brookhaven National Laboratory in New York. For a fleeting moment during this intensely energetic collision, scientists created a quark-gluon plasma. *(DOE/BNL)*

An important line of contemporary investigation involves the use of very powerful particle *accelerators* on Earth to briefly replicate the intensely energetic conditions found in the early universe. Particle physicists share the results of their very energetic particle-smashing experiments with astrophysicists and cosmologists, who are looking far out to the edges of the observable universe for telltale signatures on how the universe evolved to its present-day state from an initially intense inferno.

IMPORTANCE OF THE BIG BANG HYPOTHESIS

Many people struggle with the modern scientific concept that the universe has a finite age and that space and time actually started from nothingness some 13.7 billion years ago. Astronomy students often ask the question: Where did the big bang take place? The scientific, but unsettling, answer to this question is: nowhere, because space and time began with the big bang—the event that started the universe. Compelling scientific evidence suggests that before the big bang there was no space, there was no time, there was no universe.

Astrophysicists have three measurable signatures, which strongly support the notion that the present universe evolved from an incredibly dense, nearly featureless, extremely hot gas—just as big bang theory suggests. These scientifically measurable signatures are the expansion of the universe, the abundance of the primordial light *elements* (hydrogen, helium, and lithium), and the CMB radiation. Hubble's observation in 1929 that galaxies are generally receding (when viewed from Earth) provided scientists with their first tangible clue that the big bang theory might be correct. The big bang model also suggests that hydrogen, helium, and lithium nuclei should have fused from the very energetic collisions of protons and neutrons in the first few minutes after the big bang. (Scientists call this process big bang nucleosynthesis.) The observable universe has an incredible abundance of hydrogen and helium. Finally, within the big bang model, the early universe should have been very, very hot. Scientists regard the CMB (first detected in 1964) as the remnant heat leftover from an ancient fireball.

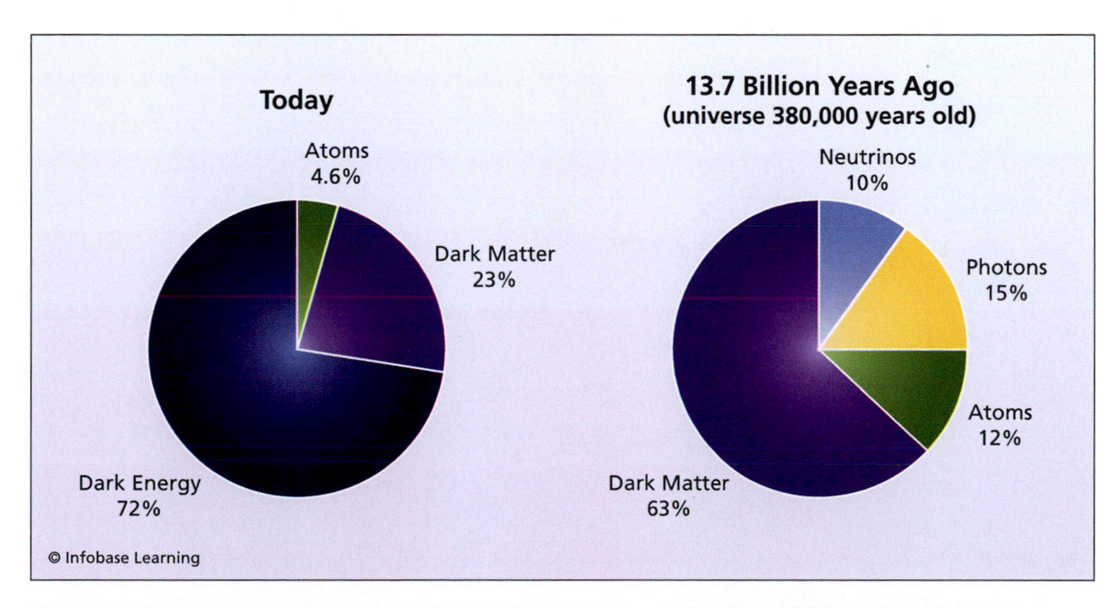

Content of the past and present universe, based upon evaluation of five years of data from NASA's *WMAP* spacecraft. (Note that *WMAP* data are only accurate to two digits, so the total appearing for today's universe is not exactly 100 percent.) *(NASA/WMAP Science Team)*

The importance of the CMB within modern cosmology cannot be understated. On March 7, 2008, NASA released the results of a five-year investigation of the oldest light in the universe. Based on a careful evaluation of *WMAP* data, scientists were able to gain special insight into the past and present content of the universe. The *WMAP* data (see the accompanying illustration on page 39) revealed that the current contents of the universe include 4.6 percent atoms—the building blocks of stars, planets, and people. In contrast, dark matter comprises 23 percent of the universe. Finally, 72 percent of the current universe is composed of dark energy, which acts like a type of matter-repulsive, anti-gravitational phenomenon.

Dark energy is distinct from dark matter and is thought to be responsible for the present-day acceleration of universal expansion. The NASA-developed illustration of the universe's contents reflects the limits inherent in the *WMAP* spacecraft's ability to estimate the amount of dark matter and dark energy by observing the CMB. Please note that *WMAP* data are accurate to two digits, so the total appearing in the content of the today's universe is not exactly 100 percent. Despite these minor limitations, the results are rather startling. The content of the current universe and the early universe (about 380,000 years after the big bang) are quite different from each other. This suggests the persistent influence of the cosmic tug-of-war between energy (radiant and dark) and matter (baryonic and nonbaryonic).

CHAPTER 3

Atomism

This chapter summarizes the atomic view of matter from ancient times up to the *standard model* used by today's scientists. The implications of thinking really small and developing the ability to manipulate matter atom by atom within the field of nanotechnology is also described.

THE ATOMIC NATURE OF MATTER

Today, almost every student in high school or college has encountered the basic scientific theory that matter consists of atoms. This generally accepted model of matter was not always prevalent in human history. The notion of atomism traces its origins back to ancient Greece.

The Greek philosopher Democritus (ca. 460–ca. 370 B.C.E.) was born in Abdera, Trace, in about 460 B.C.E. As a young man, he used his inherited wealth to travel the ancient world. He then settled down in Thrace to focus on the practice of natural philosophy. Despite the intellectual contributions of his mentor Leucippus (fifth century B.C.E.), Democritus generally receives most of the credit for being the first person to promote atomism—the idea that an atom is the smallest piece of an element and indivisible by chemical means. Long before the emergence of the scientific method, Democritus reasoned that if a person continually divides a chunk of matter into progressively smaller pieces, they eventually reach the point beyond which subdivision is no longer possible. At that point,

This 1961 stamp from Greece honors the early Greek philosopher Democritus (ca. 460–370 B.C.E.) *(Author)*

only indivisible, extremely small, tiny building blocks of matter, or atoms, remain. The modern word *atom* comes from the ancient Greek word ατομος, meaning "indivisible."

Democritus proposed that these indivisible pieces of matter (or atoms) were eternal and could not be further divided or destroyed. His atoms were also specific to the material they made up. Some types of solid matter consisted of atoms with hooks, so they could attach to one another. Other materials, such as water, consisted of large, round atoms that moved smoothly past one another. What is remarkable about the ancient Greek theory of atomism is that it tried to explain the great diversity of matter found in nature with just a few basic ideas tucked into a relatively simple theoretical framework.

Despite Democritus's clever insight, the notion of the atom as the tiniest, indivisible piece of recognizable matter languished in the backwater of human thought for more than two millennia. The main reason for this intellectual neglect was the reasoning of the influential Greek philosopher Aristotle (384–322 B.C.E.), who did not like the idea. Starting in about 340 B.C.E., Aristotle embraced and embellished the theory of matter originally proposed by Empedocles (ca. 495–435 B.C.E.). Within Aristotelian cosmology, planet Earth was regarded as the center of the universe. Aristotle speculated that everything within Earth's sphere is composed of a combination of the four basic elements: earth, air, water, and fire. Aristotle further suggested that objects made of these basic four elements were subject to change and moved in straight lines. But heavenly bodies were not subject to change and moved in circles. Aristotle also proposed that beyond Earth's sphere lies a fifth basic element, which he called the aether (αιθηρ)—a pure form of air that is not subject to change. Finally, the great Greek philosopher suggested that he could analyze all material things in terms of their matter and their form (essence). Aristotle's ideas about the nature of matter and the structure of the universe dominated thinking in Europe for centuries, until finally displaced during the Scientific Revolution.

Since Aristotle's teachings dominated Western civilization, the atomism of Leucippus and Democritus was all but abandoned. Another reason atomism did not flourish as an important concept linked to the nature of matter was the fact that the precision instruments and machines needed to effectively study atomic and nuclear phenomena began to appear only in the early part of the 20th century. Today, scientists have incredibly powerful machines that can image matter down to the atomic scale and measure intriguing processes taking place within the atomic nucleus—the very heart of matter.

During the 16th and 17th centuries, natural philosophers and scientists, such as Galileo, René Descartes (1596–1650), Robert Boyle (1627–91),

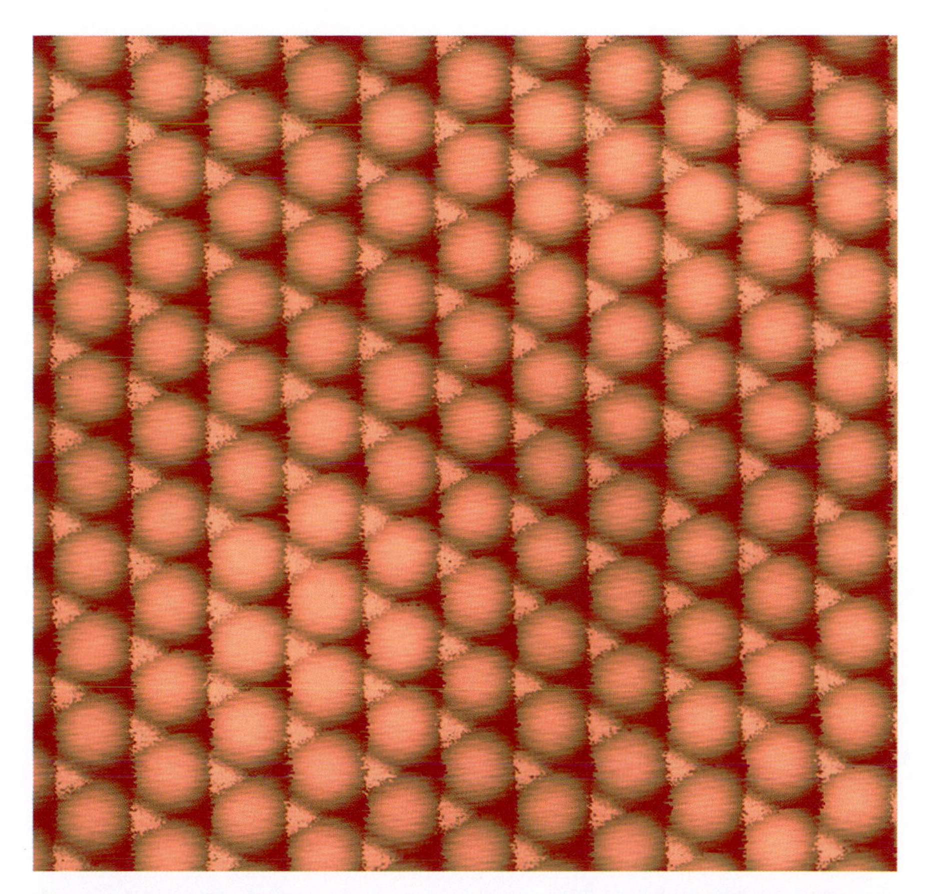

This amazing colorized image was created by the custom-built scanning tunneling microscope at the National Institute of Standards and Technology (NIST) as scientists dragged a cobalt atom across a closely packed lattice of copper atoms. *(Joseph Stroscio; Robert Celotta/NIST)*

and Sir Isaac Newton (1642–1727), all favored the view that matter was not continuous in nature but rather consisted of discrete, tiny particles or atoms. But it was not until the 19th century and the hard work of several pioneering chemists and physicists, such as John Dalton (1766–1844) and Amedeo Avogadro (1776–1856), that the concept of the atom was gradually transformed from a vague philosophical concept into a modern scientific reality.

In the 20th century, an exciting synergism occurred between the discovery of previously unimaginable nuclear phenomena and the emergence of new theories concerning the nature of matter and energy. Today, scientists use very powerful accelerators to experimentally probe inside

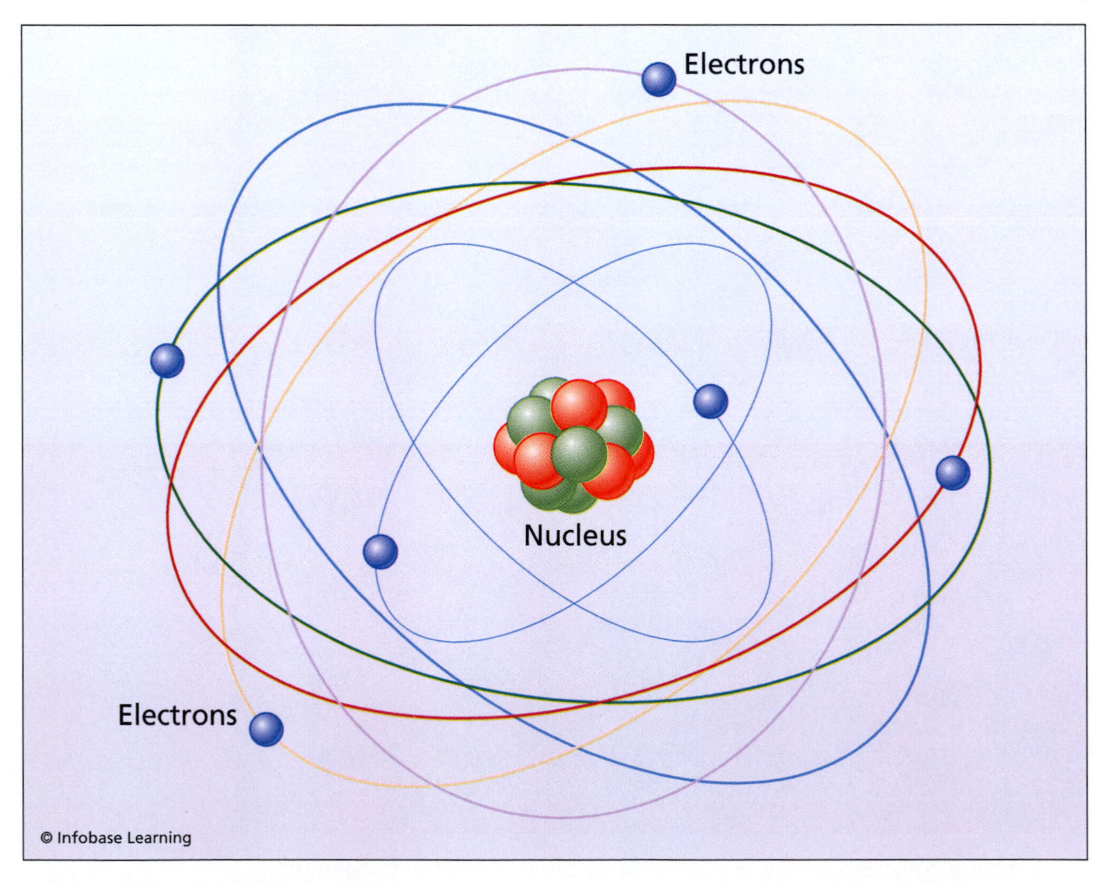

This is a simplified modern model of the atom (not drawn to scale). The dense central region, called the nucleus, consists of positively charged protons (red) and electrically neutral neutrons (green). A cloud of electrons (blue) surround the nucleus. *(Author)*

nuclear particles. They hope to validate improved theories of matter and then relate these discoveries to contemporary observations in astrophysics, which suggest the existence of puzzling phenomena such as dark matter and dark energy.

Scientists recognize that an atom is the smallest particle of matter that retains its identity as a chemical element. Atoms are indivisible by chemical means and the fundamental building blocks of all matter. The chemical elements, such as hydrogen (H), helium (He), carbon (C), iron (Fe), lead (Pb), and uranium (U), differ from one another because they consist of different types of atoms. (See appendix for a complete list of the known chemical elements.) Modern atomic theory suggests that an atom consists of a dense inner core (called the nucleus) that contains protons and neutrons and an encircling cloud of orbiting electrons.

When atoms are electrically neutral, the number of positively charged protons equals the number of negatively charged electrons. The number of protons in an atom's nucleus determines what chemical element it is; how an atom shares its negatively charged orbiting electrons determines the way that particular element physically behaves and chemically reacts. Through the phenomenon of *covalent bond*ing, for example, an atom forms physically strong links by sharing one or more of its electrons with neighboring atoms. This atomic-scale linkage ultimately manifests itself in large-scale (that is, macroscopic) material properties, such as a substance's strength and hardness.

ACCELERATORS

As nuclear physics and quantum mechanics matured in the early 20th century, other researchers began to construct machines called particle accelerators, which allowed them to hurl high-energy subatomic particles at target nuclei in an organized and somewhat controllable attempt to unlock additional information about the atomic nucleus.

The particle accelerator is one of the most important tools in nuclear science. Prior to its invention in 1932, the only known, controllable sources of particles that could induce nuclear reactions were the natural radioisotopes that emitted energetic *alpha particles*. Nuclear scientists also tried to gather fleeting peeks at energetic *nuclear reactions* by examining the avalanche of short-lived nuclear particles that rippled down toward the surface when extremely energetic cosmic rays smashed into Earth's upper atmosphere.

Modern accelerators have the particle energies and beam intensities powerful enough to let physicists probe deep into the atomic nucleus. By literally "smashing atoms," physicists can now examine the minuscule structure that lies within an individual proton or neutron. The resultant soup of quarks and gluons resembles conditions at the birth of the universe.

Common to all accelerators is the use of electric fields for the acceleration of *charged particles.* The manner in which different machines use electric fields to accelerate particles varies considerably. This section will briefly discuss the basic principles behind the operation of some of the most familiar accelerators used in nuclear science.

The most straightforward type of accelerator is the Cockcroft-Walton machine, which appeared in 1932 and opened up a new era in nuclear research. The basic device applied a potential difference between terminals. To obtain an accelerating voltage difference of more than about 200 kilovolts (kV), scientists employed one or more stages of voltage doubling circuits. The British physicist Sir John Douglas Cockcroft (1897–1967) and the Irish physicist Ernest Thomas Walton (1903–95) used their early machine to perform the first nuclear *transmutation* experiments with artificially accelerated nuclear particles (protons). For their pioneering research involving the transmutation of atomic nuclei by artificially accelerated atomic particles, they shared the Nobel Prize in physics in 1951. The Cockcroft-Walton *(direct current)* accelerator is still widely used in research.

The *radio frequency* (RF) linear accelerator repeatedly accelerates ions through relatively small potential differences, thereby avoiding problems encountered with other accelerator designs. In a linear accelerator (or linac), an ion is injected into an accelerating tube containing a number of *electrodes.* An oscillator applies a high-frequency alternating voltage between groups of electrodes. As a result, an ion traveling down the tube will be accelerated in the gap between the electrodes if the voltage is in phase. In the linear accelerator, the distance between electrodes increases along the length of the tube so that the particle being accelerated stays in phase with the voltage.

The availability of high-power microwave oscillators after World War II allowed relatively small linear accelerators to accelerate particles to relatively high energies. Today, there are a variety of large linacs, both for electron and proton acceleration, as well as several heavy-ion linacs. The Stanford Linear Accelerator (SLAC) at Stanford University in California is a 1.86-mile (3-km) long electron linac. This machine can accelerate electrons and positrons to energies of 50 GeV. Scientists are now developing superconducting radio frequency (SRF) cavities for the next generation of particle accelerators. Operating at ultracold temperatures (down

This is the Cockcroft-Walton accelerator, originally used to inject high-energy protons into the 200 million electron volt (MeV) linear accelerator (LINAC) at the Brookhaven National Laboratory (BNL) in New York. The accelerator is now retired. *(DOE/BNL)*

to −456°F [−271°C]), SRF cavities will conduct electric *current* with almost no energy loss. This means that nearly all of the electric energy input will go into accelerating the beam of particles.

The cyclotron, invented by the American physicist Ernest Orlando Lawrence (1901–58) in 1929, is the best-known and one of the most successful devices for the acceleration of ions to energies of millions of electron volts. The cyclotron, like the RF linear accelerator, achieves multiple acceleration of an ion by means of a radio frequency–generated electrical field. However, in a cyclotron a magnetic field constrains the particles to move in a spiral path. Ions are injected at the center of the *magnet* between two circular electrodes (called Dees). As a charged particle spirals outward, it gets accelerated each time it crosses the gap between the Dees. The time it takes a particle to complete an orbit is constant since the distance it travels increases at the same rate as its velocity—allowing it to stay in phase with the radio frequency signal. The favorable particle acceleration conditions break down in the cyclotron when the charged particle

Superconducting radio frequency (SRF) cavities are a key technology for the next generation of particle accelerators. Made out of superconducting material, such as niobium, the hollow cavities sit inside vessels called cryomodules and get chilled to near absolute zero. Shown here is a nine-cell 1.3-gigahertz (GHz) SRF cavity undergoing tests at the Fermi National Accelerator Laboratory (FNAL), or Fermilab, in Illinois. *(DOE/FNAL)*

or ion being accelerated reaches relativistic energies. Despite this limitation, cyclotrons remain in use throughout the world—supporting nuclear research, producing radioisotopes, and accommodating medical therapy.

THE NUCLEAR PARTICLE ZOO

From the 1930s through the early 1960s, scientists successfully developed the technical foundation for modern nuclear science. These developments included the nuclear reactor, the nuclear weapon, and many other applications of *nuclear energy.* In the process, scientists and engineers used a nuclear model of the atom that assumed the nucleus contained two basic building-block *nucleons:* protons and neutrons. This simple nuclear atom model (based on three elementary particles: proton, neutron, and electron) is still very useful in discussions concerning matter at the atomic scale.

However, as cosmic-ray research and accelerator experiments began to reveal an interesting collection of very short-lived, subnuclear particles, scientists began to wonder what was really going on within the nucleus. They pondered whether some type of interesting behavior involving neutrons and protons was taking place deep within the nucleus.

The stampede of new particles began rather innocently in 1930, when the Austrian-Swiss theoretical physicist Wolfgang Pauli (1900–58) suggested that a particle, later called the neutrino by the Italian-American physicist Enrico Fermi (1901–54), should accompany the beta decay process of a radioactive nucleus. In naming the new particle, Fermi used the Italian word meaning "little neutral one." Neutrinos were finally observed through experiments in 1956. Today, physicists regard neutrinos as particles that have almost negligible (if any) mass and travel just below (or at) the speed of light. Neutrinos are stable members of the lepton family.

Two exciting new particles joined the nuclear zoo in 1932, when the American physicist Carl David Anderson (1905–91) discovered the positron and the British physicist Sir James Chadwick (1891–1974) the neutron. Knowledge of the positron (the first observed antimatter particle) and neutron encouraged scientists to discover many other strange particles.

One of the most important early hypotheses about forces within the nucleus took place in the mid-1930s, when the Japanese physicist Hideki Yukawa (1907–81) suggested that nucleons (that is, protons and neutrons) interacted by means of an exchange force (later called the strong force). He proposed that this force involved the exchange of a hypothetical subnuclear particle called the pion. This short-lived subatomic particle was eventually discovered in 1947. The pion is a member of the meson group of particles within the hadron family.

Since Yukawa's pioneering theoretical work, remarkable advances in accelerator technology allowed nuclear scientists to discover several hundred additional particles. Virtually all of these elementary (subatomic) particles are unstable and decay with lifetimes between 10^{-6} and 10^{-23} second. Overwhelmed by this rapidly growing population of particles, nuclear scientists no longer felt comfortable treating the proton and the neutron as elementary particles.

While a detailed discussion of all these exciting advances in the physics of matter on the nuclear scale is beyond the scope of this book, the following discussion should make any visit to the nuclear particle zoo more comfortable. Physicists divide the group of known elementary particles into three families: photons, leptons, and hadrons. The basic discriminating factor involves the nature of the force by which a particular type of particle interacts with other particles. Physicists currently recognize four fundamental forces in nature: gravitation, electromagnetism, the weak force, and the strong force.

Scientists regard the photon as a stable, zero rest mass particle. Since a photon is its own antiparticle, it represents the only member the photon family. Photons interact only with charged particles, and such interactions take place via the electromagnetic force. A common example is Compton scattering by *X-rays* or *gamma rays*. Scientists consider the photon unique; they have yet to discover another particle that behaves in just this way.

The lepton family of elementary nuclear particles consists of those particles that interact by means of the weak force. Current interpretations of how the weak force works involve refinements in quantum electrodynamics (QED) theory. This theory, initially introduced by the American physicist Richard Feynman (1918–88) and others in the late 1940s, combines Maxwell's electromagnetic theory with quantum mechanics in a way that implies electrically charged particles interact by exchanging a virtual photon. (Responding to the Heisenberg *uncertainty principle,* a virtual particle exists for only an extremely short period of time.) There is good experimental verification of QED. Leptons can also exert gravitational and (if electrically charged) electromagnetic forces on other particles. Electrons (e), muons (μ), tau particles (τ), neutrinos (ν), and their antiparticles are members of this family. The electron and various types of neutrinos are stable, while other members of the family are very unstable with lifetimes of a microsecond (10^{-6} s) or less.

The hadron family contains elementary particles that interact by means of the strong force and have a complex internal structure. This family is further divided into two subclasses: mesons (which decay into leptons and photons) and baryons (which decay into protons). Hadrons may also interact by electromagnetic and gravitational forces, but the strong force dominates at short distances of 3.28×10^{-15} ft (1.0×10^{-15} m) or less. The pion (meson), neutron (baryon), and proton (baryon) are members of the hadron family, along with their respective antiparticles. Most hadrons are very short-lived with the exception of the proton and the neutron. The proton is stable, and the neutron is stable inside the nucleus but unstable outside the nucleus—exhibiting a half-life of about 12 minutes.

In the early 1960s, the American physicist Murray Gell-Mann (1929–) and others introduced quark theory to help describe the behavior of hadrons within the context of the theory of quantum chromodynamics (QCD). Quark theory suggests that hadrons are actually made up of combinations of subnuclear particles, called quarks—Gell-Mann adapted *quark* from a passage in the James Joyce comic fictional work *Finnegans Wake*. Contemporary quark theory suggests the existence of six types of quarks: called up (u), down (d), strange (s), charm (c), top (t), and bottom (b), as well as their corresponding antiquarks. Scientists first discovered quarks during experiments at the Stanford Linear Accelerator in the late 1960s and early 1970s. The top quark was the last quark to be experimentally discovered. This event took place in 1995 at the Fermi National Accelerator Laboratory.

Experiments indicate that the up (u) quark has an energy equivalent mass of 3 MeV, a charge of +2/3 (that of a proton), and a spin of 1/2. The down (d) quark has a mass of 6 MeV, a charge of –1/3, and a spin of 1/2. The charm (c) quark has a mass of 1,300 MeV, a charge of +2/3, and a spin of 1/2. The strange (s) quark has a mass of 100 MeV, a charge of –1/3, and a spin of 1/2. The top (t) quark has a mass of 175,000 MeV, a charge of +2/3, and a spin of 1/2. Finally, the bottom (b) quark has a mass of 4,300 MeV, a charge of –1/3, and a spin of 1/2.

The first family of quarks consists of up and down quarks, which are the quarks that join together to form protons and neutrons. The second family of quarks consists of strange and charm quarks, which exist only at high energies. The third family of quarks consists of top and bottom quarks. These last two quarks exist only under very high-energy conditions, as might occur briefly during accelerator experiments.

In their attempt to explain what is happening inside the atomic nucleus, scientists needed to learn more about how quarks make up protons and neutrons. Physicists now postulate that quarks have not only electromagnetic charge (in the somewhat odd fractional values mentioned above) but also an unusual, different type of charge called *color charge.* It is the force between color-charged quarks that gives rise to the strong force that holds quarks together as they form hadrons. Scientists call the carrier particles associated with the strong force, gluons (g).

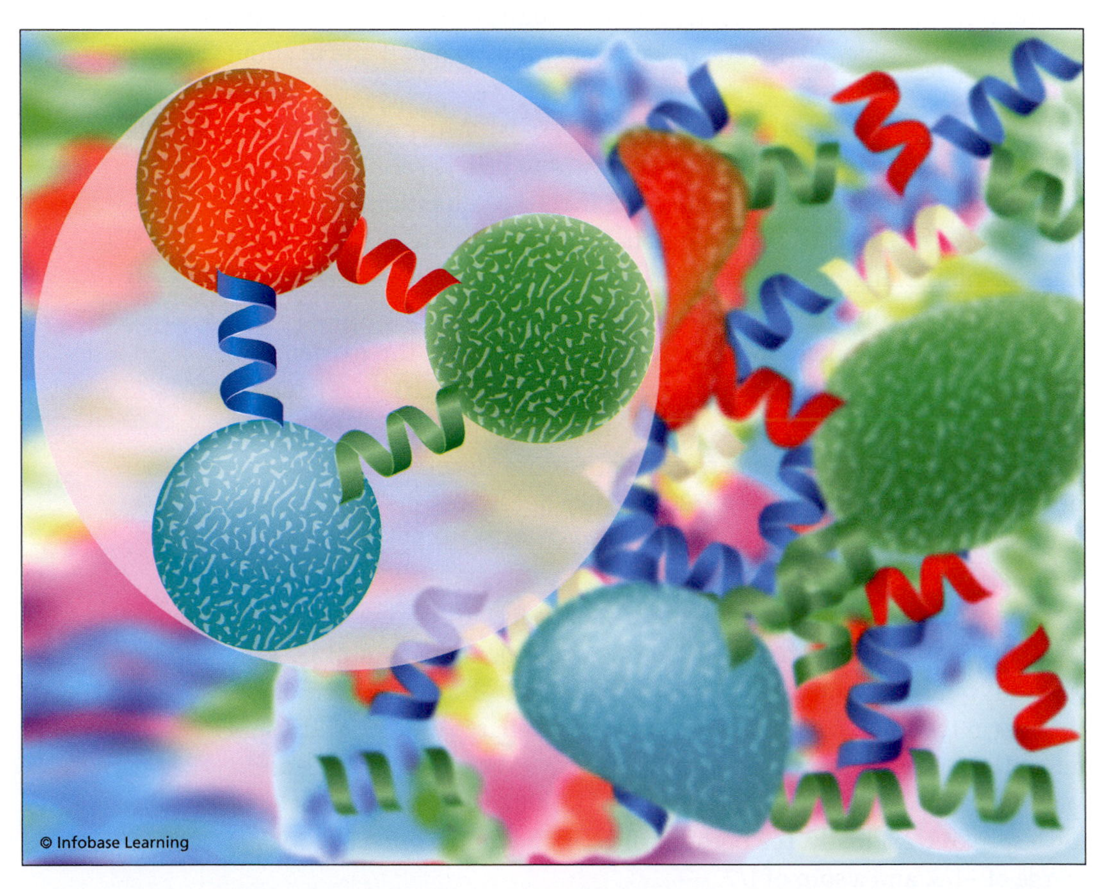

This illustration contrasts quarks within hadrons (upper left) and those within a hot, dense quark-gluon plasma (right). Protons and neutrons consist of combinations of up (u) and down (d) quarks bound by color charges carried by gluons. However, quarks and gluons are unbound and free to move independently in a hot, dense quark-gluon plasma, like the one that existed immediately after the big bang. *(Adapted from DOE/LBNL artwork)*

In the standard model of matter (discussed shortly), scientists assume that quarks and gluons are color-charged particles. Color-charged particles exchange gluons, thereby creating a very strong color-force field that binds the quarks together. Please note that the term *color charge* as used by physicists does not imply a relationship with the colors of visible light as found in the electromagnetic spectrum. Rather, physicists have arbitrarily assigned various colors to help complete the description of the quantum-level behavior of these very tiny subnuclear particles within hadrons. As a result, quarks are said to have three color charges: red, green, or blue; antiquarks three corresponding color charges: antired, antigreen, and antiblue. This additional set of quantum-level properties is an attempt by scientists to bring experimental observations into agreement with the *Pauli exclusion principle.*

Quarks constantly change their "colors" as they exchange gluons with other quarks. This means is that whenever a quark within a hadron absorbs or emits a gluon that particular quark's color must change in order to conserve color charge and maintain the bound system (that is, protons and neutrons) in a color-neutral state. Scientists also propose that quarks cannot exist individually because the color force increases as the distance between the quarks increases. The residual strong force between quarks is sufficient to hold atomic nuclei together and overcome any electrostatic repulsion between protons.

HUMAN-MADE SUPERHEAVY ELEMENTS

Scientists refer to the human-made elements with an atomic number greater than 104 as the superheavy elements (SHEs) or the transactinide elements. On a contemporary *periodic table* (see the Appendix on page 199), these elements start with rutherfordium (104) and continue up through ununoctium (Uuo) (118).

Fleeting amounts (one to several atoms) of all these elements have been created in complex laboratory experiments involving the bombardment of special transuranium target materials with high-velocity ions. Nuclear physicists hope to discover zones of stability (representing potentially longer-lived nuclei) as certain configurations of protons and neutrons are achieved in these bombardment experiments beyond element 118. Because the elements from 104 to 118 have extremely short half-lives—typically milliseconds to seconds, any type of detailed property analysis is extremely difficult.

MAKING UNUNTRIUM AND UNUNPENTIUM

In 2003, a team of American scientists from the Lawrence Livermore National Laboratory (LLNL) in California collaborated with Russian scientists from the Joint Institute for Nuclear Research (JINR) in Dubna, Russia, and their team discovered two new superheavy elements. Scientists gave elements 113 and 115 the provisional names ununtrium (Uut) and ununpentium (Uup), respectively.

The successful experiments were conducted at the JINR U400 cyclotron with the Dubna gas-filled separator between July 14 and August 10, 2003. The scientific team observed nuclear decay chains that confirmed the existence of element 115 (ununpentium) and element 113 (ununtrium). Specifically, element 113 formed as a result of the alpha decay of element 115. The experiments produced four atoms each of elements 115 and 113 through the bombardment of an americium-243 target by high-energy calcium-48 nuclei.

Scientists suggest that ununtrium's most stable *radioisotope* is Uut-284, which has a half-life of about 0.48 second. Ununtrium then undergoes alpha decay into roentgenium-280. The substance is anticipated to be a solid at room temperature, but its density is currently unknown. Scientists classify ununtrium as a metal. Ununpentium's most stable radioisotope is Uup-288, which has a half-life of about 87 milliseconds and then experiences alpha decay into Uut-284. Scientists classify ununpentium as a metal of currently unknown density. They also anticipate it would be a solid at room temperature.

Once a team of scientists using the Berkeley Laboratory's Heavy Ion Linear Accelerator (HILAC) filled in the *actinoids* (formerly actinides) in the periodic table with the identification of lawrencium (103) in 1961, scientists around the world began working to produce transactinide elements, starting with element 104. As they attempted to produce and identify these new elements, the nuclear scientists encountered several problems that made their efforts extremely difficult. Typically, only a few atoms of each suspected new element would be produced over the course of an elaborate experiment, which sometimes ran for weeks at a time. In addition, the half-lives of the artificially created transactinide elements were all extremely short, making their identification very challenging. In order to scientifically confirm that a new element has actually been created, scientists must detect and trace members of the new elements radio-

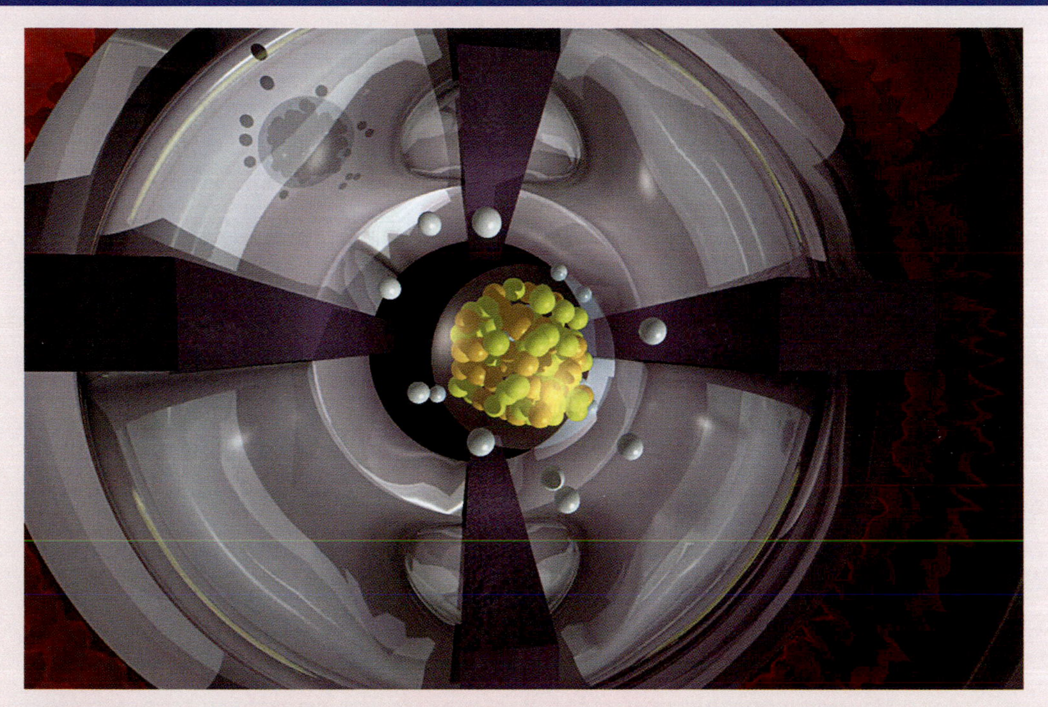

Computer-generated image depicting one of the numerous americium-243 target atoms with its nucleus of protons and neutrons surrounded by an electron cloud. In February 2004, scientists from Russia and the United States successfully bombarded americium-243 atoms with calcium-48 ions to ultimately produce several atoms of the short-lived superheavy element ununtrium (Uut) with an atomic number 113. ***(DOE/LLNL)***

active decay chain and use these radioactive emissions as the signature of the new element. The nuclear age alchemists use different combinations of target atoms and projectile atoms (ions) to produce these new elements.

In an amazing demonstration of modern materials science, researchers have been able to study the chemical properties of superheavy elements, such as rutherfordium (104) and seaborgium (106), using the advanced techniques of one-atom-at-a-time chemistry. The results, though still subject to review and discussion, suggest some continuity with the characteristic group (column) properties of lower-mass members of the periodic table.

The international quest for identifying and naming new superheavy elements in the second half of the 20th century did not take place without some controversy. Three research groups were at the center of these techno-political storms. The competitive groups involved an American team of scientists at the Lawrence Berkeley National Laboratory (LBNL) in California; a Russian team of scientists at the Joint Institute for Nuclear Research (JINR) at Dubna, Russia; and a German team of scientists at the Gesellschaft für Schwerionenforschung (GSI) in Darmstadt, Germany. Eventually, the International Union of Pure and Applied Chemistry (IUPAC) stepped in and mediated the thorny situation. In 1997, the 39th General Assembly of the IUPAC meeting in Geneva, Switzerland, approved the following names for elements 104 through 109: rutherfordium (Rf) for element 104, dubnium (Db) for element 105, seaborgium (Sg) for element 106, bohrium (Bh) for element 107, hassium (Hs) for element 108, and meitnerium (Mt) for element 109. This refereed solution did not completely please all the contentious scientific parties, but it broke an intellectual logjam that was causing a great deal of confusion within the entire international scientific community.

Fortunately, less contention and more cooperation marked the detection and naming of the remaining superheavy elements that have been observed through December 2010. The IUPAC approved the following names: darmstadtium (Ds) for element 110, roentgenium (Rg) for element 111, and copernicum Cn for element 112. The remaining transactinide elements received temporary Latin names until additional experiments verify their existence and properties. Ununtrium (Uut), meaning quite literally "one-one-three-um," is element 113; ununquadium (Uuq) is element 114; ununpentium (Uup) is element 115; ununhexium (Uuh) is element 116; ununseptium (Uus) is 117; and ununoctium (Uuo) is 118.

THE STANDARD MODEL

Scientists have developed a quantum-level model of matter they refer to as the standard model. This comprehensive model explains (reasonably well) what the material world consists of and how it holds itself together. As shown in the accompanying figure, physicists need only six quarks and six leptons to explain ordinary matter. Despite the hundreds of different particles that have been detected, all known matter particles are actually combinations of quarks and leptons. Furthermore, quarks and leptons interact by exchanging force *carrier particles*. The most familiar lepton is the electron (e), and the most familiar force carrier particle is the photon.

The standard model is a reasonably good theory and has been verified to excellent precision by numerous experiments. All of the elemen-

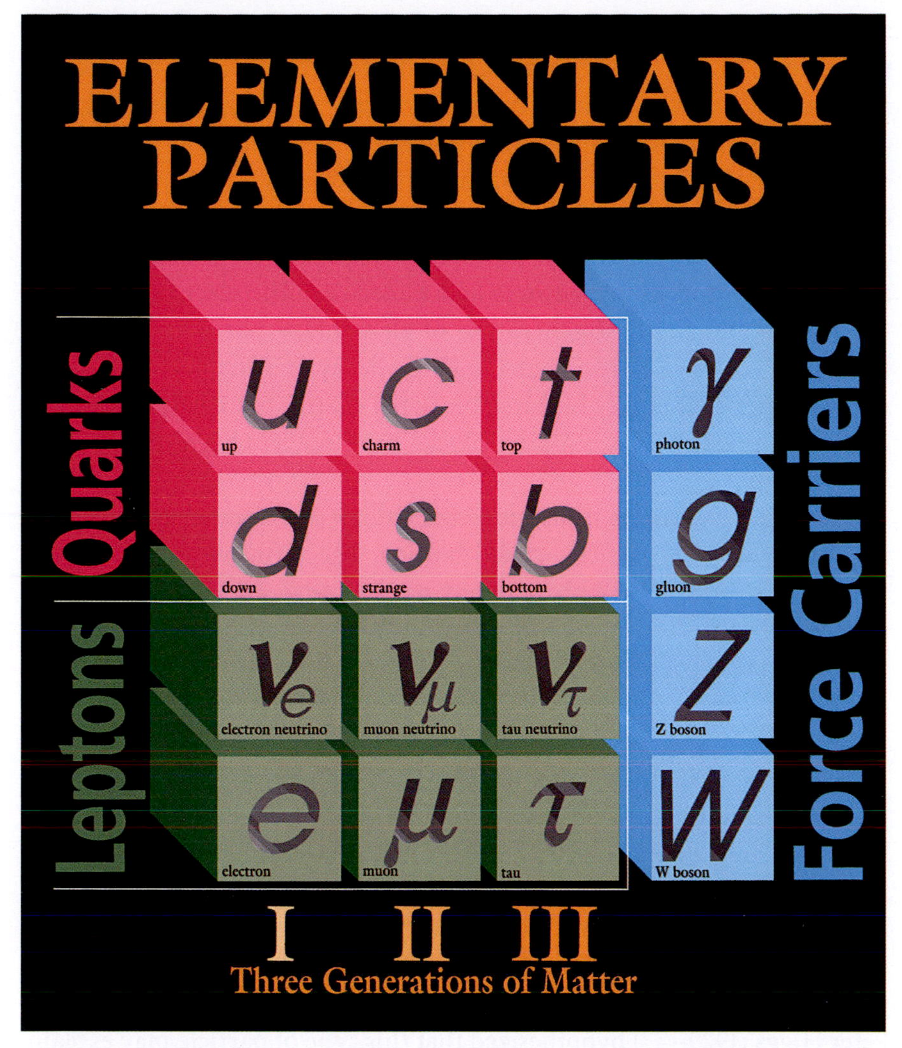

Scientists developed the standard model to explain the complex interplay between elementary particles and force carriers. *(DOE/FNAL)*

tary particles making up the standard model have been observed through experiments. But the standard model does not explain everything of interest to scientists. One obvious omission is the fact that the standard model does not include gravitation.

In the standard model, the six quarks and the six leptons are divided in pairs or generations. The lightest and most stable particles compose the first generation, while the less stable and heavier elementary particles make up the second and third generations. As presently understood by scientists, all stable (ordinary) matter in the universe consists of particles that belong to the first generation. Elementary particles in the second and third generation are heavier and have very short lifetimes, decaying quickly to the next most stable generation.

The everyday world of normal human experience involves just three of these building blocks: the up quark, the down quark, and the electron. This simple set of particles is all that nature requires to make protons and neutrons and to form atoms and molecules. The electron neutrino (ν_e) rounds out the first generation of elementary particles. Scientists have observed the electron neutrino in the decay of other particles. There may be other elementary building blocks of matter to explain dark matter, but at present such building blocks have not yet been observed in experiments.

Elementary particles transmit forces among one another by exchanging force-carrying particles called bosons. The term *boson* is the general name scientists have given to any particle with a spin of an integral number (that is, 0, 1, 2, etc.) of quantum units of angular momentum. Carrier particles of all interactions are bosons. Mesons are also regarded as bosons. The term honors the Indian physicist Satyendra Nath Bose (1894–1974).

As represented in the figure on page 57, the photon *(γ)* carries the electromagnetic force and transmits light. The gluon *(g)* mediates the strong force and binds quarks together. The *W* and *Z* bosons represent the weak force and facilitate the decay of heavier (more energetic) particles into lower mass (less energetic) ones. The only fundamental particle predicted by the standard model that has not yet been observed is a hypothetical particle called the Higgs boson. In 1964, the British theoretical physicist Peter Higgs (1929–) hypothesized that this type of particle may explain why certain elementary particles (such as quarks and electrons) have mass and other particles (such as photons) do not. If found by research scientists this century, the Higgs boson (sometimes called the God particle) could play a major role in refining the standard model and shedding additional light on the nature of matter at the quantum level. If nature does not provide scientists with a Higgs boson, then they will need to postulate other forces and particles to explain the origin of mass and to preserve the interactive components of the standard model they have already verified by experiments.

Finally, the force of gravity is not yet included in the standard model. Some physicists suggest that gravitational force may be associated with

THINKING SMALL

On December 29, 1959, the American physicist and Nobel laureate Richard P. Feynman delivered an inspiring lecture entitled "There's Plenty of Room at the Bottom." Feynman presented this public talk at the California Institute of Technology (Caltech). In the course of his discussions, Feynman suggested that it would be possible, using late 1950s technology, to write an enormous quantity of information, such as the entire content of a multivolume encyclopedia, in a tiny space equivalent to the head of a pin. He also speculated about the impact that micro- and nano-scaled machines might have.

This illustration compares in size the microbot used at the RoboCup 2009 nanosoccer competition by the team from Switzerland's ETH Zurich to the head of a fruit fly. The tiny robot, which is operated under a microscope, is just 300 micrometers in length—or slightly larger than a dust mite. ***(NIST/ETH Zurich)***

Clearly, fields such as medicine and information technology (built upon microelectronics) would be revolutionized if these (at the time) hypothetical tiny devices allowed engineers and scientists to manipulate individual atoms and arrange these building blocks of matter into useful devices. As Feynman spoke that December evening, his predicted revolution in microelectronics was actually well under way—sparked by the development of the integrated circuit (IC).

An integrated circuit (IC) is an electronic circuit that includes transistors, resistors, capacitors, and their interconnections—all fabricated on a very small piece of semiconductor material (usually referred to as a chip). The American electrical engineer Jack S. Kirby (1923–2005) invented the integrated circuit while he was working on electronic component miniaturization during the summer of 1958 at Texas Instruments. Although Kirby's initial device was quite crude (by today's standards), it proved to be a groundbreaking innovation that paved the way for the truly miniaturized electronics packages, which now define

(continues)

(continued)

the digital age. Kirby shared the 2000 Nobel Prize in physics "for his part in the invention of the integrated circuit."

As sometimes happens in science and engineering, two individuals independently come up with the same innovative idea at about the same time. In January 1959, another American engineer, Robert Norton Noyce (1927–90), while working for a California company named Fairchild Semiconductor, independently duplicated Kirby's feat. In 1971, Noyce, who was then the president and chief executive officer of another California company called Intel, developed the world's first microprocessor. Today, both Kirby and Noyce are recognized as having independently invented the integrated circuit.

The category of an integrated circuit, such as LSI and VLSI, refers to the level of integration and denotes the number of transistors on a chip. Using one common (yet arbitrary) standard, engineers say a chip has small-scale integration (SSI) if it contains less than 10 transistors, medium-scale integration if it contains between 10 and 100 transistors, large-scale integration (LSI) if it contains between 100 and 1,000 transistors, and very large-scale integration (VLSI) if it contains more than 1,000 transistors.

another hypothetical particle called the graviton. Another major challenge facing scientists in this century is to develop a quantum formulation of gravitation that encircles and supports the standard model. The harmonious blending of general relativity (which describes gravitation on a cosmic scale) and quantum mechanics (which describes the behavior matter on the atomic scale) would represent another incredible milestone in humankind's search for the meaning of substance.

To help face that challenge, scientists and engineers have designed extremely powerful particle accelerators. Two examples are the Tevatron at the Fermi National Accelerator Laboratory in Illinois and the Large Hadron Collider (LHC), which is the most powerful proton-antiproton accelerator in the world, at the European Organization for Nuclear Research (organizational acronym CERN) on the Franco-Swiss border near Geneva, Switzerland. The Tevatron is the most powerful particle accelerator in the United States. The machine can accelerate beams of protons and antiprotons to 99.99999954 percent of the speed of light around a four-mile (6.4-km) circumference. Scientists used the Tevatron to discover the top quark and to place constraints on the mass of the hypothetical, elusive Higgs boson.

The LHC at CERN is the largest, most complex, and most powerful particle accelerator ever built. The LHC can create almost a billion proton-proton collisions per second. The enormous accelerator operates in a circular tunnel almost 17 miles (27.4 km) in circumference about 330 feet (100 m) underground, between France's Jura Mountains and Switzerland's Lake Geneva. In March 2010, the LHC collided protons at a center-of-mass energy of 7 trillion electron volts (7.5 TeV) per beam. This is seven times higher than achievable at Fermilab's Tevatron. Scientists anticipate that such very-high-energy proton collisions will yield fascinating new discoveries about the nature of matter and the physical universe. Some of postulated breakthroughs include new information regarding the origins of mass and the hypothetical Higgs boson, the characteristics of dark matter, possible hidden symmetries of the universe, and the existence of extra dimensions of space (see chapter 9).

LILLIPUTIAN WORLD OF NANOTECHNOLOGY

This section introduces the amazing, extremely small world of nanotechnology. Ask 10 different scientists or engineers to define the term *nanotechnology* and most likely they will provide 10 different answers. Here, *nanotechnology* refers to the manufacturing and application of nano-scale machines, electronic devices, and chemical or biological systems—all of which have characteristic dimensions on the order of one to 100 nanometers.

Optimistic scientists and engineers view nanotechnology as ushering in an incredible new age in materials science, during which they acquire an atom-by-atom understanding of functional matter. Pessimists suggest that manipulating individual atoms, while demonstrated on a laboratory scale, may not be achievable on sufficiently large scales to result in useful products and applications. The pessimists further warn about the possible risks of legions of misbehaving nano-scale devices running amok, once they get into the environment or within the human body. As with all major technical breakthroughs in human history, what actually occurs will most likely fall between these two significantly divergent projections. The purpose here is to briefly discuss ongoing efforts in nanotechnology—a combined scientific and industrial effort involving the atomic-level manipulation of matter that can significantly influence the trajectory of civilization.

Feynman's futuristic speculations about micro-machines stimulated interest in the creation of very tiny devices that could perform useful functions at the near atomic scale. Within a decade or so, engineers began creating such micro-size devices. Today, scientists and engineers construct MEMS (*m*icro-*e*lectro-*m*echanical *s*ystem) devices with dimensions on the order of one to 10 micrometers. Science historians generally regard Feynman's lecture as the most convenient milestone marking the beginning of nanotechnology.

According to the National Nanotechnology Initiative (NNI) established by the United States government in 2001, nanotechnology is the understanding and control of matter at dimensions between approximately one and 100 nanometers. A nanometer (nm) is just one-billionth of a meter (that is, 1×10^{-9} m). Therefore, nano-scale research refers to science, engineering, and technology performed at the level of atoms and molecules. The practice of nanotechnology involves imaging, measuring, modeling, and manipulating matter within this incredibly tiny physical realm—a Lilliputian world that is certainly well beyond day-to-day human experience. A single gold atom is about one-third of a nanometer in diameter. Ten hydrogen atoms placed in a row (side by side) would span about one nanometer. By comparison, the width of a human hair ranges from about 50,000 to 150,000 nanometers, and a sheet of paper is about 100,000 nanometers thick. The accompanying figures illustrate the scale of some typical things found in nature and some very small human-made objects.

Scientists are beginning to appreciate that the nano-scale region represents the physical scale at which the fundamental properties of materials and systems are established. For example, how atoms are arranged in nanostructures determines such physical attributes as a material's melting temperature, magnetic properties, and capacity to store charge. Biologists and medical researchers recognize that molecular biology functions for the most part at the nano-scale level.

Scientists and engineers know a great deal about the physical properties and behavior of matter on the macroscopic scale. They also understand the properties and behavior of individual atoms and molecules when these are considered as isolated, quantum-level systems. But they cannot easily predict the nano-scale behavior of a material by simply extrapolating either its macroscopic (bulk) properties or its quantum-level properties. Rather, at the nano-scale level, collections of atoms exhibit important differences in physical properties and behavior that cannot be easily explained using traditional theories and models.

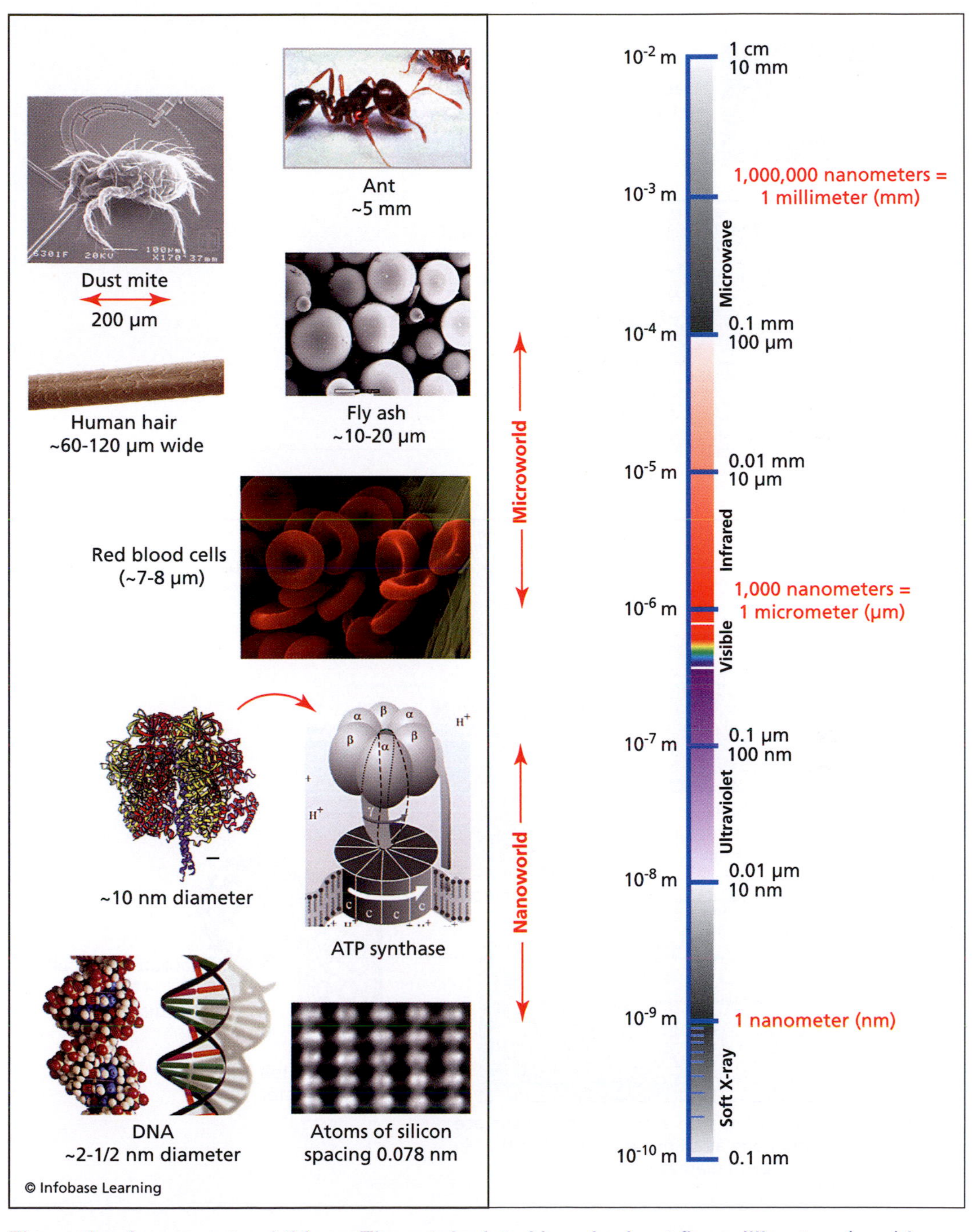

The scale of some natural things. The ant depicted here is about five millimeters (mm) in length, while silicon atoms are spaced about 0.078 nanometer (nm) apart. The vertical scale helps differentiate what scientists mean by the terms *microscale* and *nanoscale*. *(Adapted from DOE artwork)*

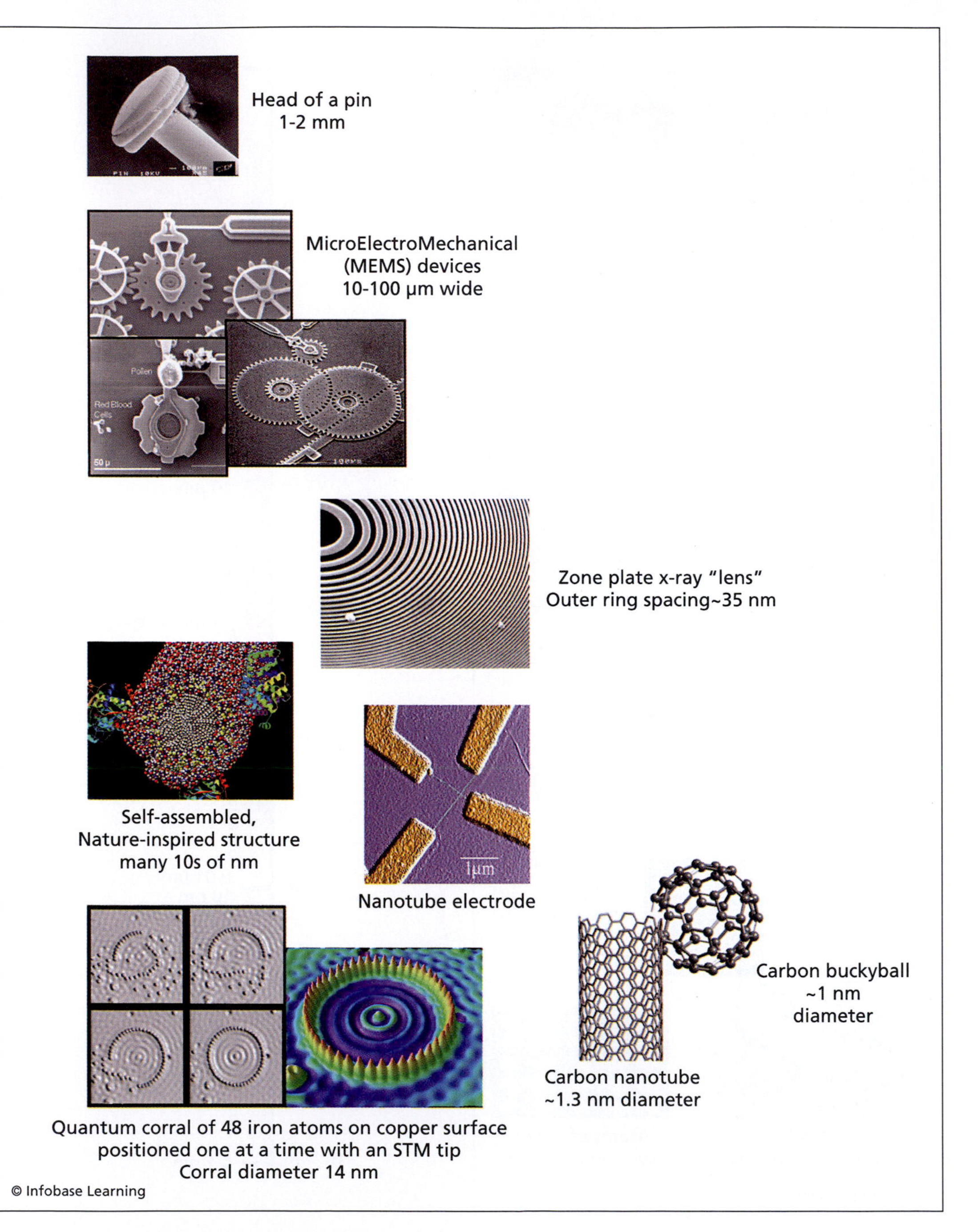

The scale of some human-made things. Objects depicted range from the head of a pin (about 1 to 2 mm diameter) to a carbon buckyball (about 1 nm diameter). *(Adapted from DOE artwork)*

Scientists define *nanoparticle* as referring to a very tiny chunk of matter (that is, less than 100 nanometer on a side). This term is less specific than the term *molecule* because it is dimension-specific rather than being related to the chemical composition of the tiny piece of matter. Almost any material substance can be transformed into nanoparticles. Researchers are discovering that nanoparticles of a particular material often behave quite differently from large amounts of the same substance. Some of these differences in physical properties or behavior result from the continuous modification of a material's characteristics as the sample size changes. Other differences seem to represent phenomena associated with the quantum-level behavior of matter, such as wavelike transport and quantum-size confinement. Scientists have also observed that beginning at the molecular level new physical and chemical properties emerge as cooperative interactions and interfacial phenomena start dominating the behavior of nano-scale molecular structures and complexes.

One of the promises of nanotechnology research is for scientists to develop the ability to understand, design, and control these interesting properties. When this occurs, they will then be able to construct functional molecular assemblies—starting a new revolution in materials science. By performing nano-scale research, scientists will develop a basic understanding of the physical, chemical, and biological behavior of matter at these very tiny dimensions. Once improved understanding of material behavior at the nano-scale occurs, scientists and engineers can set about the task of crafting customized new materials and introducing functional atomic-level devices. Researchers project potential benefits influencing medicine, electronics, biotechnology, agriculture, transportation, energy production, environmental protection, and many other fields. As part of ongoing nano-scale research activities, scientists and engineers are learning how molecules organize and assemble themselves into complex nano-scale systems and then begin to function. In the future, they will start constructing a variety of customized quantum devices capable of serving human needs.

BUCKYBALLS

A significant milestone in the development of nanotechnology occurred in 1985, when the researchers Richard E. Smalley (1943–2005), Robert F. Curl, Jr. (1933–), and Sir Harold W. Kroto (1939–) discovered an amazing molecule, which consisted of 60-linked carbon atoms (C_{60}).

NASA's *Spitzer Space Telescope* detected buckyballs in interstellar space. This artist's rendering shows buckyballs floating in interstellar space, near a region of current star formation. The intriguing, miniature soccer ball–shaped carbon molecule is the largest molecule ever discovered floating between the stars. *(NASA/JPL-Caltech/T. Pyle [SSC/Caltech])*

Smalley named these distinctive clusters of carbon atoms buckyballs in honor of the famous American architect Richard Buckminster Fuller (1895–1983), who promoted the use of geodesic domes. Scientists had known for centuries that carbon consisted of two allotropes—namely, diamond and graphite. So their collaborative discovery of another carbon allotrope quite literally rocked the world of chemistry.

Buckyballs are very hard to break apart. When slammed against a solid object or squeezed, they bounce back. The cluster of 60 carbon atoms is especially stable. It has a hollow, icosahedral structure in which the bonds between the carbon atoms resemble the patterns on a soccer ball. Smalley, Curl, and Kroto shared the 1996 Nobel Prize in chemistry for their "discovery of carbon atoms found in the form of a ball."

Identification of this new allotrope of carbon sparked broad interest in the chemistry of an entire family of hollow carbon structures, now referred to collectively as fullerenes. Formed when vaporized carbon condenses in an inert atmosphere, fullerenes include a wide range of shapes and sizes—including nanotubes, which are of interest in electronics and hydrogen storage. Nanotubes are cylindrically shaped tubes of carbon atoms about one nanometer in diameter that are stronger than steel and can conduct electricity. The walls of such nanotubes have the same soccer-ball-like structure as buckyballs, but they come rolled up into long tubes. Fullerenes represent a major research area in nanotechnology. Since they come in many variations, highly versatile fullerenes promise many potential applications. Researchers think that fullerene structures can be manipulated to produce superconducting salts, new *catalysts,* new three-dimensional *polymers,* or biologically active *compounds.*

CHAPTER 4

Very Hot Matter

This chapter discusses plasma, the *state of matter* involving a highly ionized gas in which the number of positively charged nuclei (ions) is approximately equal to the number of free electrons. Scientists often refer to plasma as the fourth state of matter. Natural plasmas are the most common form of visible matter in the observable universe. They are found in stars and throughout interstellar space. On Earth, human-made plasmas are commonly used in artificial light sources. Scientists are also manipulating plasmas in experimental thermonuclear fusion reactors.

PLASMA—FOURTH STATE OF MATTER

When a gas is heated to a sufficiently high temperature, all the atoms become ionized. Scientists refer to a highly ionized gas as plasma. Although a plasma contains free positive ions and free negative electrons, the numbers of positive and negative electrical charges balance. Therefore, when viewed as a whole, a plasma is electrically neutral. Because of the presence of the electrically charged particles, plasmas have a number of interesting *physical properties.* As discussed later in this chapter, scientists are attempting to use some of these unique properties, such as how plasmas respond to imposed magnetic fields, in their efforts to achieve controlled thermonuclear fusion.

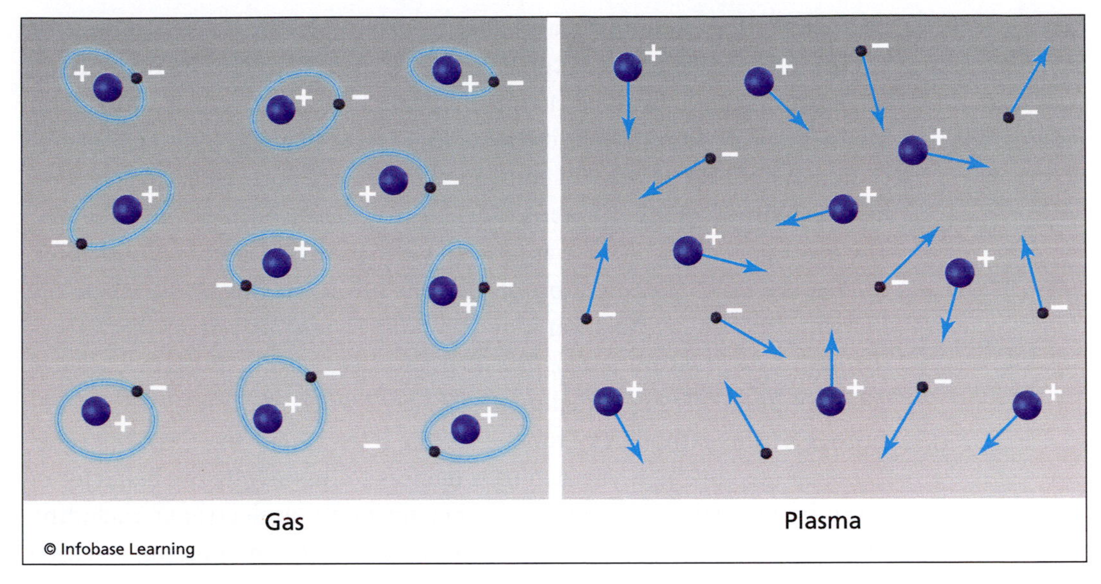

As a gas gets heated to a very high temperature, electrons become free of their atomic nuclei and a plasma is formed. When viewed as a whole, a plasma is electrically neutral. *(Based on DOE artwork)*

At high temperatures, an atom's negatively charged orbiting electrons become separated from the positively charged atomic nucleus. Scientists call this separation process ionization, and they refer to the positively charged nuclei as *ions*. Because of the electric charges carried by electrons and ions, scientists can use magnetic fields to confine a plasma. In the absence of a magnetic field, a plasma (or assembly) of charged particles in a cylindrical containment vessel will move in straight lines in random directions. With nothing to restrict their motion, these charged particles will quickly strike the wall of the containment vessel. Such collisions tend to cool the plasma down. This is an undesirable side effect for scientists trying to use high-temperature plasmas to promote controlled nuclear fusion.

When scientists apply a uniform magnetic field on a plasma, the motion of charged particles is confined and they follow spiral (helical) paths about the magnetic field lines. Positively charged particles spiral in one direction; negatively charged particles spiral in the other direction. As a result, the spiraling charged particles are not free to travel at random across magnetic field lines. The use of a magnetic field to confine a high-temperature plasma also prevents the charged particles from striking the wall of the containment vessel. Although the mathematics describing particle motion in uniform, and nonuniform magnetic fields can be quite complex, the basic physical concepts are fairly easy to comprehend. For a uniform magnetic field, each charged particle is "attached" to the line of magnetic force along which it travels in a spiral (helical) path of constant radius. As part of decades of controlled nuclear fusion research, scientists have examined many different types of magnetic confinement schemes. The most promising of these approaches are discussed later in the chapter.

The scientific investigation of plasmas began in the 1830s, when the British scientist Michael Faraday (1791–1867) sent electrical discharges through low-pressure gases. The British scientist Sir William Crookes (1832–1919) greatly expanded upon Faraday's work. He invented the fore-runner of the cathode ray tube, a device later called the Crookes tube. Crookes used his high-vacuum tube devices to investigate the behavior of cathode rays (later recognized by other scientists as electrons), including their deflection by an applied magnetic field. These pioneering research activities made him one of the first scientists to investigate plasma. In 1879, Crookes suggested that the "radiant matter" in discharge tubes was a fourth state of matter.

While studying electrical discharges in gases, the American physical chemist and Nobel laureate Irving Langmuir (1881–1957) coined the term

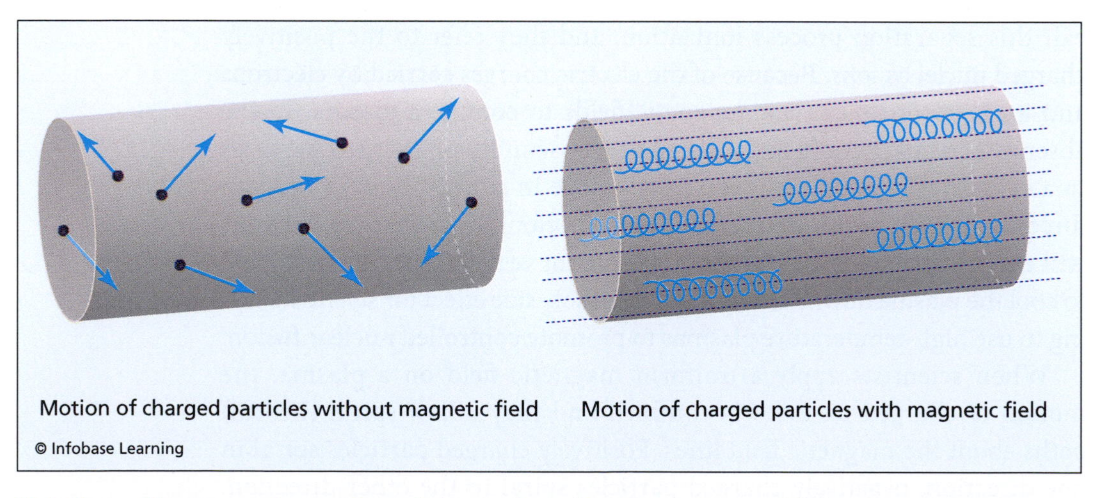

This figure illustrates the motion of charged particles in a plasma. Without a magnetic field, the charged particles in a plasma move randomly in straight lines, often striking the wall of the containment vessel. Scientists use a magnetic field to confine and direct the motion of the charged particles in a high-temperature plasma and to prevent them from striking the wall of the containment vessel. *(Based on DOE artwork)*

A noncommissioned officer in the U.S. Air Force uses a plasma arc-cutting torch inside the welding shop at an air base in Germany. *(U.S. Air Force)*

plasma to describe the ionized gases. In his notes, Langmuir suggested that ionized gases "reminded him of blood plasma," as found in biology and medicine. In 1924, he developed the diagnostic technique and instrument (an electrostatic probe now called a Langmuir probe) that allowed scientists to measure the (electron) temperature and density of a plasma. As a spin-off of his pioneering work with plasmas, Langmuir developed atomic hydrogen welding—the first application of plasma in welding.

The Swedish scientist Hannes Olof Gösta Alfvén (1908–95) promoted breakthroughs in plasma physics by developing a comprehensive explanation of the behavior of plasma in the presence of magnetic fields. Alfvén's pioneering work formed the foundation of magnetohydrodynamics (MHD)—the branch of physics that studies the interactions between a magnetic field and a conducting fluid, such as a plasma.

TYPES AND CHARACTERISTICS OF PLASMAS

Plasma physics proved of great interest to astronomers in the 20th century. More than 99 percent of matter in the observable universe is in the plasma state. Stars are so hot that their large quantities of matter can exist only as plasma. (Stars are discussed in the next chapter.) Interstellar space

Artist's rendering of the solar wind interacting with Earth's magnetosphere (Not drawn to scale) *(NASA)*

is also filled with plasma. Closer to Earth, the solar wind is a variable stream of plasma (that is electrons, protons, alpha particles, and other atomic nuclei) that flows continuously outward from the Sun through interplanetary space.

Naturally occurring plasmas are not common on Earth. Lightning is the most familiar and observable natural example. At any given moment, there are about 2,000 thunderstorms occurring across the globe. Meteorologists estimate that there are between 40 and 100 lightning flashes taking place in Earth's atmosphere each second. It is the flashes of lightning associated with these storms that creates the thunder. Although scientists still have much to discover about lightning, they generally describe the phenomenon as a gigantic flash of static *electricity.* Lightning can be as hot as 54,000°F (30,000°C)—a temperature that is five times hotter than the Sun's surface (photosphere). When lightning occurs, it heats and ionizes the *air* surrounding its channel to the same incredible temperature in a fraction of a second. Thunder is the acoustic shock wave that results from the extreme heat suddenly generated by a lightning flash.

Earth's ionosphere and the aurora are other examples of natural plasma related to humans' home planet. The ionosphere is that portion of Earth's upper atmosphere, which extends from about 31 miles (50 km) to 625 miles (1,000 km) altitude. It contains ions and free electrons in sufficient quantities to reflect radio waves. The aurora are the visible glow in Earth's upper atmosphere (ionosphere) caused by the interaction of the planet's magnetosphere and particles from the Sun (that is, the solar wind). The aurora borealis (or northern lights) and the aurora australis (or southern lights) are visible manifestations of the dynamic behavior of Earth's magnetosphere. At high altitudes, disturbances in Earth's geomagnetic field accelerate trapped particles into the upper atmosphere, where they excite nitrogen molecules (red emissions) and oxygen atoms (red and green emissions). Assisted by scientific spacecraft, scientists have also observed auroras on the planets Jupiter, Saturn, Uranus, and Neptune.

Various artificial lighting systems represent the most prevalent human-made plasmas found on Earth. High-intensity arc lamps and fluorescent lamps are the two most commonly encountered plasma-based lighting devices. Typically, engineers use plasma to directly produce the light given off by a high-intensity arc lamp. Chemical elements within the plasma determine the characteristic color of the high-intensity light.

The aurora borealis (or northern lights) blaze across the night sky above Bear Lake, Alaska, in January 2005. The auroras are the result of solar wind particles colliding with and ionizing gases in Earth's upper atmosphere. Native peoples of the Arctic created different legends about the northern lights—such as that they were the souls of animals dancing in the sky. *(U.S. Air Force; Senior Airman Joshua Strang photographer)*

High-intensity arc lamps provide outdoor and security lighting in public areas, at commercial and industrial facilities, and (on occasion) near individual residences. Commercially available, high-intensity discharge (HID) lamps provide high efficiency and long service life. In an HID discharge lamp, electricity arcs between two electrodes, creating an intensely bright light in a special arc tube. Mercury, sodium, or metal halide gases

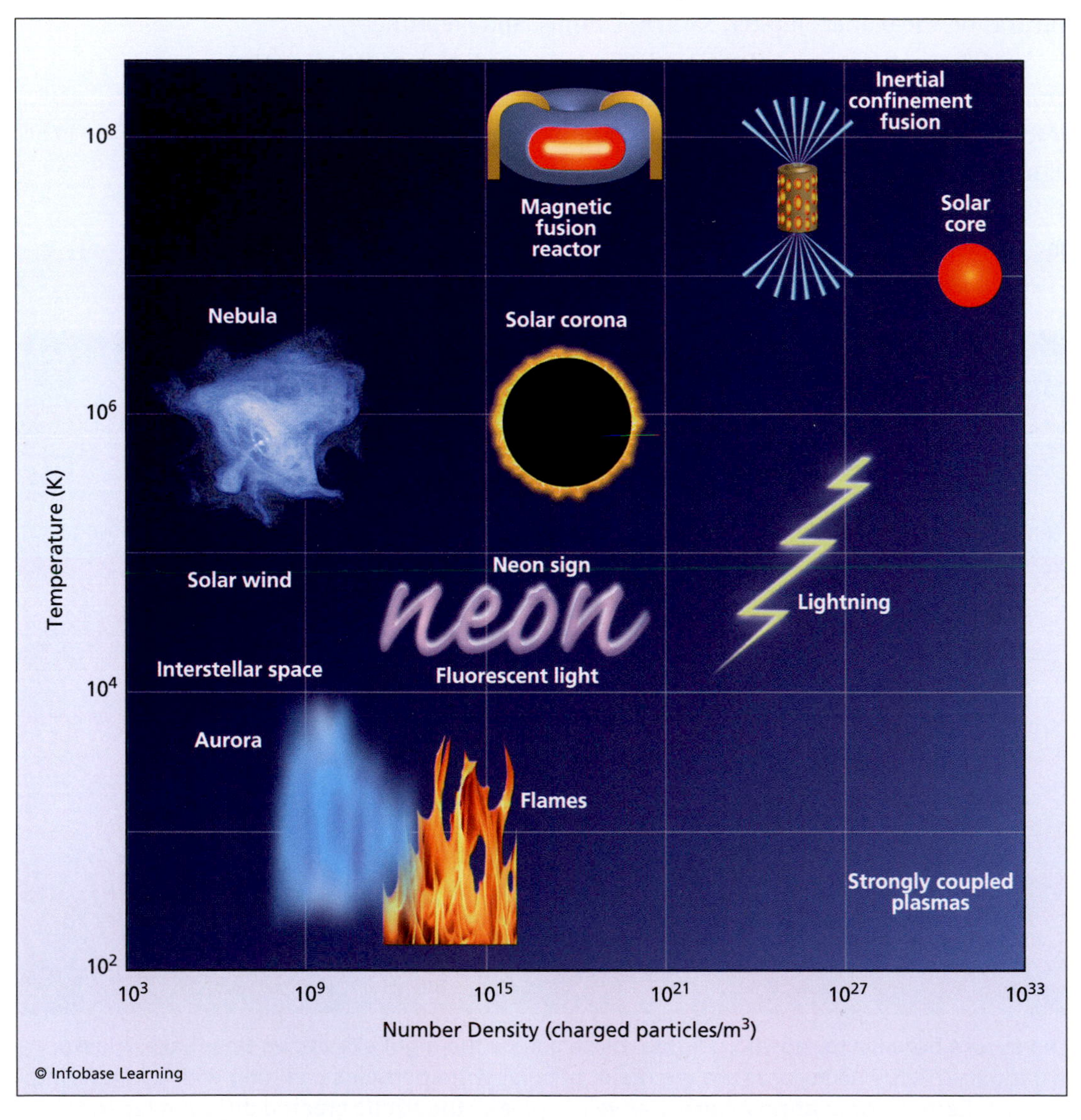

This illustration depicts the range of plasma temperatures and particle (number) densities for a variety of human-made and natural plasma systems. *(Based on DOE/LANL artwork)*

serve as the conductor. Because they can take up to 10 minutes to produce light when first turned on, HID lamps are most suitable for lighting applications where they stay on for hours at a time.

Fluorescent lamps are very common within homes, offices, factories, and commercial buildings. Generally, a plasma within the central portion of the fluorescent lamp stimulates the phosphor (white coating) that lines the inside portions of the lamp's external surface. Once stimulated, the phosphor emits the soft white light people generally associated with fluorescent lamps. From an engineering perspective, commercially available fluorescent lamps are essentially gas-discharge devices that use electric energy to excite mercury vapor. Excited atoms in the mercury plasma emit ultraviolet light, which then stimulates the phosphor material, causing it to emit visible light.

There are many other applications of human-made plasmas ranging from flat-panel displays to plasma-based thrusters for spacecraft propulsion and attitude-control subsystems. The colorful neon lights found in signs and advertising displays represent still other human-made plasma devices. Technical visionaries suggest that an exciting future application of human-made plasmas will be in controlled thermonuclear fusion reactors.

The accompanying figure shows where many plasma systems (natural and human-made) occur when compared with respect to their charged particle densities and plasma temperatures. Plasma temperatures and number densities range from relatively cool and tenuous (such as associated with auroras) to very hot and dense (such as the core or central region of a star).

In plasmas, the separation of electrons and ions produce electric fields, and the motion of ions and electrons produce both electric and magnetic fields. As a result, plasmas are rampant with instabilities, nonlinearities, and chaotic phenomena, making them much more difficult to model than solids, liquids, or gases. Any further detailed, analytical discussion of the plasma state lies beyond the scope of this book. However, several basic classifications used by plasma physicists may prove helpful without being mathematically burdensome.

Scientists define an ideal plasma as one in which Coulomb collisions are negligible. Like-charged particles experience Coulomb repulsion; while unlike (oppositely)-charged particles experience Coulomb attraction. When Coulomb collisions become significant, they regard the plasma as nonideal or strongly coupled. Scientists recognize that at low-particle

MAGNETOPLASMADYNAMIC THRUSTER

The magnetoplasmadynamic thruster (MPD) is an advanced electric propulsion device capable of operating with a wide range of propellants in both pulsed and steady-state modes. MPD thrusters are well suited for orbit transfer and spacecraft maneuvering applications. In an MPD thruster device, the current flowing from the *cathode* to the *anode* sets up a ring-shaped magnetic field. This magnetic field then pushes against the plasma in the arc. As propellant, such as argon, flows through the arc plasma, it is ionized and forced away by the magnetic field. A thrusting force therefore is created by the interaction of an electrical current and a magnetic field.

densities and low-plasma temperature, they can reasonably approximate a system of partially ionized plasma as a *mixture* of ideal gases, consisting of electrons, ions, and atoms. Under these physical conditions, the particles generally move in straight lines and only occasionally collide (interact) with one another. As the plasma's density increases, the particles start interacting more with one another. Eventually, the density of the plasma gets sufficiently high, and scientists say the plasma has become strongly coupled or nonideal. The interior (core) of stars, fusion plasmas, and the solar wind are examples of fully ionized plasmas. Earth's ionosphere is an example of a partially ionized plasma.

NUCLEAR FUSION

Plasma research activities experienced a major shift of emphasis in the 1940s. Under government sponsorship, scientists in the United States began to concentrate their plasma research activities on the creation and control of nuclear fusion reactions. For example, Project Matterhorn was the code name given to a classified government effort at Princeton University to develop the science and technology to control thermonuclear reactions. Other nuclear fusion research took place at the Los Alamos National Laboratory in New Mexico and later at the Lawrence Livermore National Laboratory in California as part of an urgent national effort to develop the so-called hydrogen bomb in the early 1950s.

The American astrophysicist Lyman Spitzer, Jr. (1914–97) made major contributions in the areas of stellar dynamics, plasma physics, and thermonuclear fusion. In 1951, Spitzer founded the Princeton Plasma Physics Laboratory (PPPL) under Project Matterhorn—a classified effort sponsored by the U.S. Atomic Energy Commission. With declassification, the laboratory emerged as Princeton University's pioneering program in controlled thermonuclear reaction research. Spitzer promoted efforts to harness nuclear fusion as a clean source of energy and remained the laboratory's director until 1967. In the early 1950s, as part of the original Project Matterhorn effort, Spitzer proposed a pioneering magnetic confinement device called the stellarator. This device used an arrangement of twisted magnetic fields to contain very hot plasma.

Halfway around the world, in various nuclear research laboratories of the former Soviet Union, Russian physicists such as Nobel laureates Igor Evgenievich Tamm (1895–1971) and Andrei Dmitrievich Sakharov (1921–89) were also exploring plasma physics as related to both thermonuclear weapons and controlled nuclear fusion systems. The tokamak concept for the magnetic confinement of nuclear fusion reactions emerged from these activities. (The word *tokamak* is an acronym for the Russian phrase "toroidal chamber with magnetic coil.") A tokamak system uses strong magnetic fields to contain hot plasma within a hollow, doughnut-shaped (torroidal) tube.

As part of several international meetings in the late 1950s and early 1960s dealing with the peaceful use of nuclear energy, scientists around the world began to openly share their research work concerning controlled thermonuclear fusion. After more than five decades of research, many problems still remain with developing and maintaining hot plasma capable of supporting nuclear fusion reactions. While *nuclear weapons* scientists (in certain countries) have been able to harness nuclear fusion reactions for a very brief instant to achieve extremely energetic explosions, nuclear energy engineers are still unable to harness nuclear fusion in a controlled, dependable manner so as to yield significant outputs of useful energy.

In nuclear fusion, lighter atomic nuclei are joined together, or fused, to form a heavier nucleus. For example, the fusion of deuterium (${}^{2}_{1}D$) with tritium (${}^{3}_{1}T$) results in the formation of a helium nucleus and a neutron. Deuterium is the stable, nonradioactive isotope of hydrogen that contains one proton and one neutron in its nucleus. Tritium is the naturally occurring radioisotope of hydrogen with two neutrons and one proton in the nucleus. With a half-life of just 12.3 years, the tritium used in thermonuclear weapons and in controlled fusion research is human-made in *nuclear reactors* through the neutron irradiation of the *isotope* lithium-6.

The total mass of the fusion *products* (a neutron and an alpha particle) is less than the total mass of the *reactants* (that is, the original deuterium and tritium nuclei), a tiny amount of mass disappears in D-T fusion, and the equivalent amount of energy is released in accordance with Albert

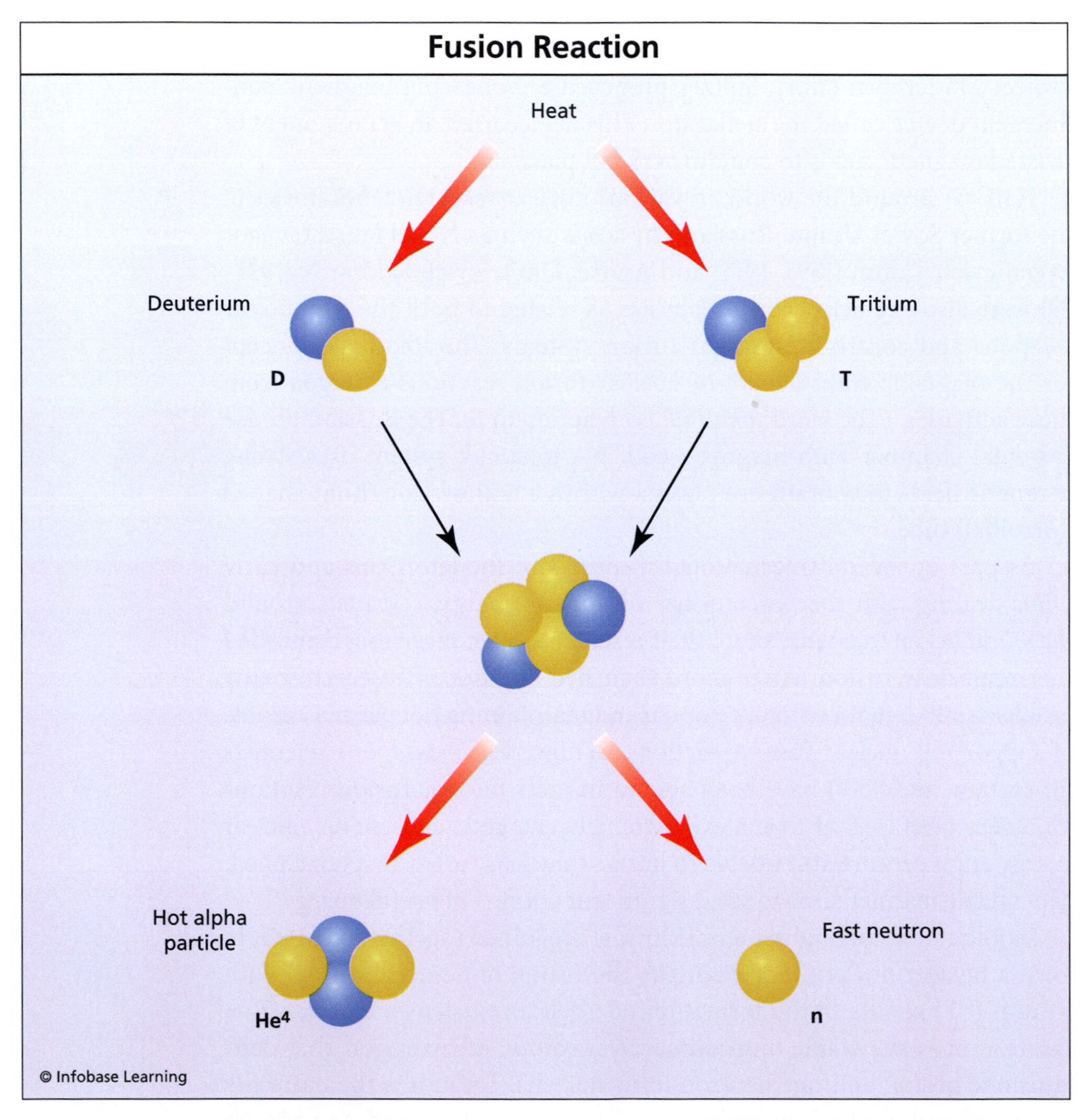

The basic deuterium (D)–tritium (T) nuclear reaction, the most readily attainable fusion process on Earth. Each reaction releases about 17.6 MeV, which appears as the kinetic energies of the two product particles. The emerging neutron carries away about 14.1 MeV; the energetic (hot) alpha particle about 3.5 MeV. *(DOE)*

Einstein's mass-energy equivalence formula: $E = \Delta mc^2$. This fusion energy then appears as the *kinetic* (motion) *energy* of the reaction products. Nuclear scientists discovered that when isotopes of elements lighter than iron fuse together, some fusion energy is liberated. However, energy must be added to any fusion reaction involving elements heavier than iron.

The Sun is Earth's oldest source of energy and the mainstay of most terrestrial life. The energy of the Sun and other stars comes from thermonuclear fusion reactions. Fusion reactions brought about by means of very high temperatures are called thermonuclear reactions. The actual temperature required to join, or fuse, two atomic nuclei depends on the nuclei and the particular fusion reaction involved. The two nuclei being joined must have enough energy to overcome the Coulomb repulsive force between like-signed (here, positive and positive) electric charges. In stellar interiors, fusion occurs at temperatures of tens of millions of rankines (kelvins) or more. For example, solar physicists estimate that the temperature at the center of the Sun is about 27×10^6 R (15×10^6 K). When scientists try to harness controlled thermonuclear reactions (CTRs) on Earth, they must use techniques involving reaction temperatures starting at about 180 million rankines (100 million kelvins).

In the late 1950s, the British physicist John David Lawson (1923–2008) specified the theoretical conditions involving the plasma particle density (n) and confinement time (τ) needed in a self-sustaining nuclear fusion system. Lawson based his specifications (later called the Lawson criterion) on the requirement that in a self-sustaining nuclear fusion system the reacting nuclei must be confined long enough to produce sufficient recoverable energy by fusion to compensate for the amount of energy supplied to heat the plasma. The Lawson criterion is expressed as the product $n\tau$, where n is the plasma density (typically given in particles per cubic centimeter) and τ is the time (typically given in seconds) that a magnetic field can confine or contain the reacting plasma. Lawson suggested that a human-engineered nuclear fusion reactor must have an $n\tau$ in excess of 10^{14} for plasmas involving the deuterium-tritium (D-T) nuclear fusion reaction, and in excess of 10^{16} for plasmas involving the deuterium-deuterium (D-D) nuclear fusion reaction.

Scientists and engineers note that there are a range of conditions over which the Lawson criterion can be satisfied, since it involves the product of particle density and confinement time. For example, if a deuterium-tritium plasma at a temperature of 180×10^6 R (100×10^6 K) has a particle density of 10^{14} particles/cm^3, then the confinement time needed for a self-sustaining fusion system exceeds one second; but if the particle density is 10^{16}, then the confinement time need be only longer than 0.01 second. Similarly, a hot plasma of deuterium at 900×10^6 R (500×10^6 K) with particle density of 10^{14} particles/cm^3 requires a confinement time in excess of 100 seconds to achieve self-sustaining D-D fusion. However, if the particle density increases to 10^{16}, then the confinement time must just exceed one second.

Because nuclei have positive charges, they normally repel one another. The higher the temperature, the faster the nuclei are moving. When they collide at very high speeds, they can overcome the Coulomb repulsion barrier, and the nuclei can fuse. The challenge faced by scientists and engineers in developing a useful fusion power system is to develop a device that can heat the fuel to sufficiently high temperature and then confine the hot plasma for long enough so that more energy is released through nuclear fusion reactions than is expended in heating the plasma.

To release energy at a practical level for use in electric power generation, a gaseous mixture of deuterium-tritium fuel must be heated to about 180×10^6 R (100×10^6 K). In an operating (future) fusion reactor, part of the energy generated will help to maintain the plasma temperature as fresh quantities of deuterium and tritium are introduced. In contemporary magnetic confinement fusion experiments, insufficient fusion energy is released to maintain the plasma temperature. As a result, the experimental devices operate in short pulses, and the plasma must be reheated in every pulse.

Because plasma conducts electricity, scientists can heat plasma by passing a current through it. They refer to the process as resistive (or ohmic) heating. As the plasma's temperature increases, however, its resistance to the flow of electricity decreases and the ohmic heating process becomes less effective. To heat the plasma even more, scientists next use another technique called neutral beam injection. They introduce high-energy atoms into the ohmically-heated, magnetically-confined plasma. The newly injected atoms are immediately ionized and become trapped within the magnetic field. By transferring a portion of their kinetic energy to the plasma particles through repeated collisions, the injected atoms increase the plasma temperature. Scientists also use radio frequency waves to heat the charged particles in the plasma and to drive the plasma current.

Through these various techniques, scientists are able to heat plasmas to extremely high temperatures. However, they are still faced with the daunting technical challenge of confining the deuterium and tritium plasma under such extreme conditions. Scientists use magnetic confinement techniques to contain the hot plasma particles within a working volume and to prevent them from striking the walls of the containment vessel. Scientists now anticipate that successful fusion systems with their D-T plasmas heated to temperatures of 180 million R (100 million K) need to have the Lawson criterion product (nτ) greater than 3×10^{14} particles-second/cm^3.

At present, there are immense technical difficulties preventing the effective use of controlled fusion as a terrestrial or space energy source. The key problem is that the fusion gas mixture must be heated to extremely high temperatures and then confined for a sufficiently long period of time

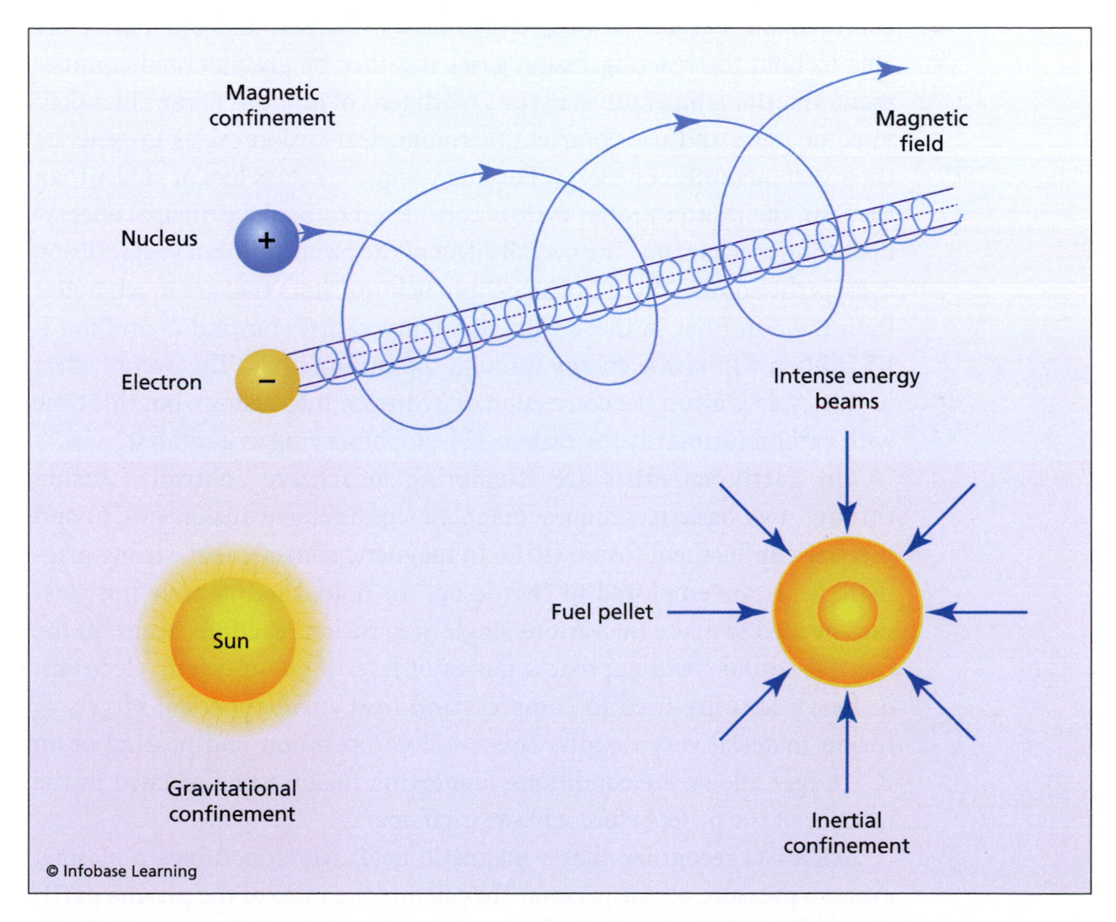

This illustration shows the three different approaches to keeping the hot fuel (plasma) for fusion reaction confined despite the immense Coloumb force (repulsion of like-charged particles) that wants to drive reacting nuclei away from one another. *(Adapted from DOE artwork)*

for a useful number of fusion reactions to occur. As previously mentioned, a deuterium-tritium (D-T) gas mixture must be heated to a minimum temperature of about 180×10^6 R (100×10^6 K)—and nuclear physicists consider this the "easiest" controlled fusion reaction for them to achieve. The minimum temperature for a deuterium-deuterium (D-D) reaction is about 900×10^6 R (500×10^6 K). At such extreme temperatures, any physical material used to confine the reacting fusion gases would disintegrate, and the vaporized wall materials would then "cool" the fusion gas mixture, quenching the reaction.

There are three general approaches to confining these hot fusion gases, or plasmas: gravitational confinement, magnetic confinement, and inertial confinement. Because of their large masses, the Sun and other stars are able to hold the reacting fusion gases together by gravitational confinement. Interior temperatures in stars reach tens of millions of rankines (kelvins) or more and use complete thermonuclear-fusion cycles to generate their vast quantities of energy. For main-sequence stars like or cooler than the Sun, the proton-proton cycle is considered to be the principal energy-liberating mechanism. The overall effect of the proton-proton stellar fusion cycle is the conversion of hydrogen into helium. Stars that are much hotter than the Sun (that is, those with core temperatures beyond 27 million R [15 million K]) release energy through the carbon cycle. The overall effect of this cycle is again the conversion of hydrogen into helium, but this time with carbon (primarily the carbon-12 isotope) serving as a catalyst.

On Earth, scientists are attempting to achieve controlled fusion through two basic techniques: magnetic-confinement fusion (MCF) and inertial-confinement fusion (ICF). In magnetic confinement, strong magnetic fields are employed to "bottle up," or hold, the intensely hot plasmas needed to make the various single-step fusion reactions occur. In the inertial-confinement approach, pulses of laser light, energetic electrons, or heavy ions are used to compress and heat small spherical targets of fusion material very rapidly. This rapid compression and heating of an ICF target allows the conditions supporting fusion to be reached in the interior of the pellet before it blows itself apart.

Scientists recognize that a magnetic field, which confines a plasma, exerts a pressure on the plasma. They define the ratio of the plasma particle pressure to the magnetic field pressure by a figure of merit called beta (β). For magnetically confined plasma, the value of beta can range from unity (one) down to almost zero. When beta is one, the plasma pressure is

ITER PROJECT

The main objective of the international fusion project called ITER (Latin for "the way") is to demonstrate the scientific and technological feasibility of fusion energy, using an advanced magnetic-confinement device known as a tokamak. The project is being designed and built by the ITER partners: the European Union, India, Japan, the People's Republic of China, the Republic of Korea, the Russian Federation, and the United States.

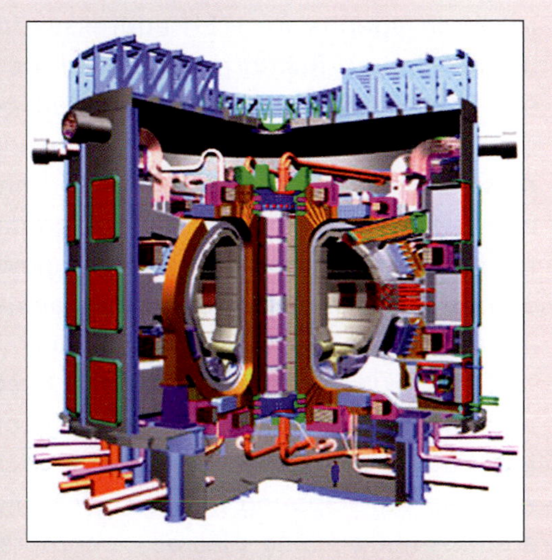

Cutaway artist's rendering of the ITER, an advanced tokamak-type magnetic confinement device now under construction at an international controlled fusion research facility at Cadarache in southeastern France. ***(DOE)***

The international team of scientists and engineers regards ITER as the premier scientific tool for exploring and testing expectations for plasma behavior in the fusion-burning plasma regime, wherein the deuterium-tritium fusion process itself provides the dominant heat source to sustain the plasma temperature at about 180,000,000 R (100,000,000 K). This research project is intended to provide the scientific basis and plasma control tools needed to move the world's energy-hungry population toward a controlled thermonuclear fusion energy economy. The ITER device is now under construction at Cadarache in southeastern France, with the European Union serving as the host party. First plasma in the toroidal magnetic-confinement device is anticipated before 2020.

As currently anticipated, a successful ITER device would produce 500 megawatts of fusion power for 500 seconds or longer during each burst (shot) of the fusion experiment, with a repetition period of roughly 2,000 seconds. In contrast, one of ITER's technical predecessors, the Tokamak Fusion Test Reactor (TFTR) at the Princeton Plasma Physics Laboratory produced a maximum of 11 megawatts for only one-third of a second. The TFTR was shut down in 1997.

(continues)

(continued)

Under the auspices of the U.S. Department of Energy, the American research agenda for ITER addresses four fundamental questions relevant to understanding fusion plasmas and the surrounding environment. First, can a self-heated fusion plasma be created, controlled, and sustained? Second, how does the large size of the plasma required for a fusion power plant affect the plasma's confinement, stability, and energy? Third, can the tokamak confinement concept be extended to the continuous, self-sustaining regime required for future controlled fusion-based power plants? Finally, what materials and components are suitable for the plasma containment vessel and its surrounding structures in a fusion power plant?

The U.S. ITER Project Office is hosted by Oak Ridge National Laboratory with the Princeton Plasma Physics Laboratory and the Savannah River National Laboratory as partner laboratories. Scientists from universities and industry are also participating in the ITER project's efforts to create and understand a sustained burning plasma. A comprehensive understanding of the burning plasma state is needed to confidently extrapolate plasma behavior and related technology beyond ITER to a fusion power plant.

equal to the magnetic field pressure; when beta is almost zero, the plasma pressure is very low. For magnetic confinement of a plasma to occur, beta can never exceed unity; otherwise, the plasma would immediately escape from the magnetic field. The smaller the value of beta (below unity), the greater the penetration of the magnetic field lines into the plasma.

There are two basic categories of magnetic confinement systems: closed confinement and open confinement. In general, closed magnetic confinement systems form a torus, or doughnut-shaped, hollow chamber. With this toroidal arrangement, the magnetic field lines are closed and charged particles cannot escape by traveling along the magnetic field lines. However, the confined plasma can drift to the chamber walls, especially if the confining magnetic field is not uniform.

In the open-ended approach to magnetic confinement, the plasma is confined in a cylindrical vessel and the magnetic lines run parallel to the length of the cylinder. Magnetic mirrors (regions of higher magnetic field strength) reflect the charged particles back to regions of lower magnetic field strength, thereby preventing the plasma from escaping from the ends of the cylinder.

In the 1970s, researchers started experimenting with powerful laser beams in an effort to compress and heat small quantities of deuterium and tritium to the point of fusion. Scientists called the technique inertial-confinement fusion (ICF). In the direct drive approach to ICF, teams of physicists and engineers carefully focused powerful beams of laser light on a small spherical pellet that contained microgram quantities of deuterium and tritium.

The rapid heating caused by the laser beams made the outer layer of the tiny target explode. Consistent with laws of physics (especially Newton's third law of motion, the action-reaction principle), as the outer surface material uniformly blew off, the inner material reacted and drove inward, causing compression of the fuel inside the capsule and the formation of a shock wave. This caused further heating of the fuel in the very center of the capsule and resulted in a self-sustaining thermonuclear burn known to scientists as ignition. The fusion burn then propagated outward very

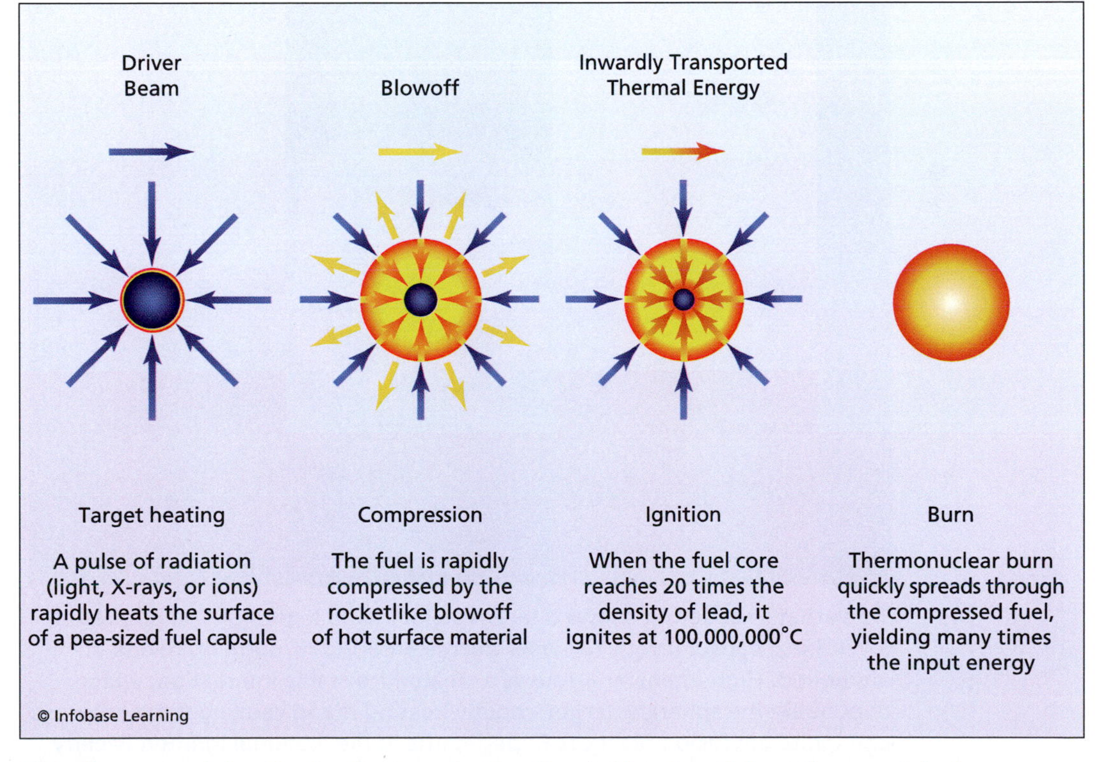

This figure explains the basic steps in the direct drive inertial confinement fusion process. *(DOE)*

rapidly, through the cooler outer regions of the capsule. The fusion reactions propagated outward much more rapidly than the mechanical expansion of the capsule material. Throughout this process, the hot plasma was confined by the inertia of its own mass—giving rise to the term *inertial-confinement fusion.*

In direct drive ICF, the small pellet containing deuterium and tritium is compressed to a very high density. The nuclei collision rate then increases and the fuel burns (that is, experiences thermonuclear reactions) before it disassembles. Inertia holds the fuel together during this very brief (about 10^{-11}-second) period of thermonuclear burn. Scientists can compress the thermonuclear fuel pellet using laser beams, ion-beams, or X-rays. When scientists use laser light to achieve ICF, the process is often called laser fusion. If the beams shine directly on the pellet, the target is referred to as direct-drive.

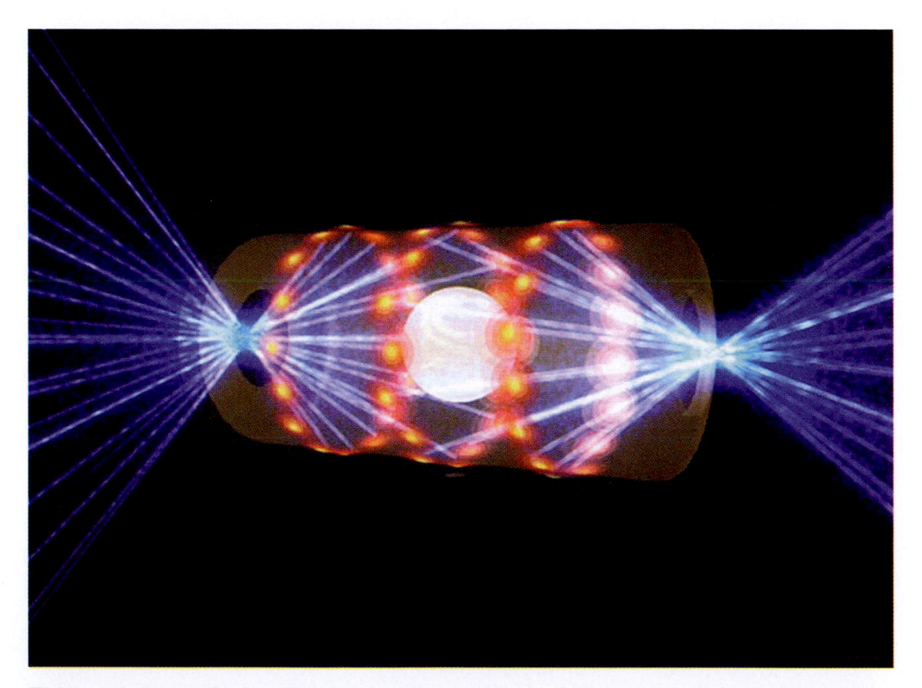

This artist's rendering shows a target pellet inside a gold-plated hohlraum (cylindrical capsule) with 192 laser beams entering through openings at either end. High-intensity X-rays are created from this interaction, which then bathe the spherical target, rapidly heating it and causing it to compress and ignite the fusion reaction. Experiments at the National Ignition Facility are a critical step in developing commercially viable fusion power plants, based on such inertial confinement fusion processes. *(DOE/LLNL)*

There is another approach to ICF, called the indirect drive method. At the National Ignition Facility (NIF) in California, scientists at the Lawrence Livermore National Laboratory are using the beams from 192 powerful lasers to heat the inner walls of a cylindrically shaped gold cavity called a hohlraum. As shown in the accompanying illustration, the hohlraum contains a fuel tiny capsule of deuterium and tritium. As the laser light enters the hohlraum, it creates a very hot plasma, which radiates a uniform bath of soft X-rays. The X-rays then rapidly heat the outer surface of the fuel pellet, causing a high-speed blow-off (ablation) of the surface material. Reacting to the blow-off, the fuel capsule then implodes in the same way, as if the capsule had been directly hit by multiple beams of high-intensity laser light. The symmetrical compression of the fuel capsule by the hohlraum's X-ray radiation forms a central hot spot, where the fusion process then occurs. The plasma ignites, and the compressed fuel undergoes a thermonuclear burn before it can disassemble.

Scientists anticipate that experiments at the National Ignition Facility will release more energy from nuclear fusion than the amount of energy they input to power the lasers that promoted the ICF process. Some projections suggest an energy output to input gain ranging between 10 and 100. Successful research at NIF would represent a very significant step toward making fusion energy viable in commercial power plants later in this century.

On November 2, 2010, scientists at the National Ignition Facility set a world record for energy delivered to an ICF target. That date, the NIF team successfully fired 1.3 megajoules of ultraviolet laser light into a cryogenically cooled hohlraum, containing a surrogate fusion target known as a symmetry capsule, or symcap. This test was the highest energy laser shot to date at the facility and was the first test of hohlraum temperatures and capsule symmetry under the extreme conditions designed to produce fusion ignition and energy gain. Preliminary analyses indicated that the hohlraum absorbed nearly 90 percent of the incoming laser energy and reached a peak radiation temperature of approximately 6×10^6 R (3.3×10^6 K)—the highest X-ray drive energy ever achieved in an indirect drive ignition target.

THERMONUCLEAR WEAPONS

In sharp contrast to previous and current scientific attempts at controlled nuclear fusion for power and propulsion applications, since the early 1950s nuclear weapon designers have been able to harness certain fusion reactions in advanced nuclear weapon systems called thermonuclear devices. In these types of nuclear explosives, the energy of a fission device is used to create the conditions necessary to achieve (for a brief moment) a significant number of fusion reactions of either the deuterium-tritium (D-T) or deuterium-deuterium (D-D) kind. Very powerful modern thermonuclear

The 11-megaton yield ROMEO thermonuclear test detonated by the United States on a barge in the Pacific Ocean at Bikini Atoll on March 27, 1954 (local time). This successful test of a high-yield thermonuclear device was part of Operation Castle. *(DOE)*

weapons have been designed and demonstrated with total explosive yields from a few hundred kilotons (kT) to the multimegaton (MT) range. A megaton yield device has the explosive equivalence of 1 million tons of the chemical high explosive trinitrotoluene (TNT).

As part of the cold war nuclear arms race, the United States detonated its first experimental thermonuclear device, called MIKE, during Operation Ivy on Enewetak Atoll in the Pacific Ocean. This surface detonation took place on October 31, 1952, and produced an explosive yield of about 10.4 megatons (MT). On August 12, 1953, the former Soviet Union conducted its first thermonuclear weapon test. Atmospheric testing by both countries continued over the next several years. For example, the United States conducted Operation Castle at the Pacific Proving Ground in 1954. During this particular exercise, a number of very successful high-yield thermonuclear tests were conducted, the results of which enabled American nuclear bomb designers to rapidly transform bulky test devices into efficient, compact, aircraft-transportable weapon systems of mass destruction.

Modern nuclear weapons produce their nuclear explosions by initiating and sustaining *nuclear chain reactions* in highly compressed material, which can undergo both fission and fusion reactions. Contemporary strategic weapons in the U.S. nuclear arsenal use a nuclear package with two assemblies: the primary assembly (which is used as the initial source of energy) and the secondary assembly (which provides additional explosive release). The primary assembly contains a central core, called the pit, which is surrounded by a layer of chemical *high explosive.* The pit is typically composed of plutonium-239 and/or highly enriched uranium (HEU) and other materials. Radiation from the explosion of the primary assembly is contained and used to transfer energy to compress and ignite a physically separate secondary component containing thermonuclear fuel. The secondary assembly is typically composed of lithium deuteride (LiD), uranium, and other materials. As the secondary assembly implodes, the lithium (in the isotopic form lithium-6) is converted to tritium (T) by neutron interactions. The tritium product, in turn, undergoes fusion with the deuterium (D) to create a thermonuclear explosion.

CHAPTER 5

Life Cycles of Stars

This chapter focuses on stars—the self-luminous balls of plasma that liberate energy through thermonuclear reactions deep within their cores. Stars experience a fascinating evolutionary life cycle from birth in a giant interstellar cloud of dust and gas to death as a white dwarf, neutron star, or black hole. Stars are the basic units of collective mass in the observable universe. Through their energy-liberating lives and sometimes spectacularly explosive deaths, stars manufacture all the chemical elements beyond hydrogen and helium—thereby making life possible. As contemporary astronomical discoveries indicate, a variety of planets, some experiencing extreme physical conditions, serve as companions to other stars within the Milky Way galaxy.

STARS AND THE EARLY UNIVERSE

At the beginning of the matter-dominated era, some three minutes (or 180 seconds) following the big bang, the expanding universe cooled to a temperature of 1.8×10^9 R (1×10^9 K) and low-mass nuclei, such as deuterium, helium-3, helium-4, and (a trace quantity of) lithium-7 began to form as a result of very energetic collisions in which the reacting particles joined or fused. Once the temperature of the expanding universe dropped below 1.8×10^9 R (1×10^9 K), environmental conditions were no longer favorable to support the additional fusion of low-mass nuclei and the process of big bang nucleosynthesis ceased. Scientists note that the observed amount of helium in the universe agrees with the theoretical predictions of big bang nucleosynthesis.

Later, when the expanding universe reached an age of about 380,000 years, the temperature had dropped to about 5,400 R (3,000 K), allowing electrons and nuclei to combine and form hydrogen and helium atoms. Interstellar space is still filled with the remnants of these primordial hydrogen and helium atoms. Within the period of atom formation, the

(continues on page 94)

This is an artist's rendering of how the early universe (less than 1 billion years after the big bang) must have looked when it went through a very rapid onset of star formation, converting primordial hydrogen into a myriad of stars at an unprecedented rate. Regions filled with newborn stars began to glow intensely. Brilliant new stars began to illuminate the cosmic Dark Ages. The most massive of these early stars self-detonated as supernovas, creating and spreading chemical elements throughout the fledgling universe. Analysis of *Hubble Space Telescope* data supports the hypothesis that the universe's first stars (called population III stars) appeared in an eruption of star formation, rather than at a gradual pace. *(NASA, K. Lanzetta [SUNY], and Adolf Schaller [STScI])*

THE EARLY UNIVERSE

The illustration on page 93 provides a brief summary of the history of the universe from the big bang, which took place about 13.7 billion years ago, to the present day. Emphasis is placed on two important cosmic epochs: hydrogen reionization and helium reionization.

In late 2010, a team of astronomers announced their discovery of an overheated early universe. They had used data from the Cosmic Origins Spectrograph (COS) instrument aboard NASA's *Hubble Space Telescope (HST).* The young universe went through an initial "heat wave" more than 13 billion years ago, when radiant energy from early massive stars ionized cold interstellar hydrogen produced in the first three minutes following the big bang. Ionization takes place when electrons are removed from atoms, producing free electrons and positively charged atomic nuclei called ions. Scientists call this cosmic episode the hydrogen reionization epoch, because the primordial hydrogen nuclei had originally been in an ionized state from shortly after the big bang and remained so up until the universe was about 380,000 years old.

The helium reionization epoch took place from 11.7 to about 11.3 billion years ago, when the early universe burned off a fog of primordial (big bang created) helium. As the *Hubble Space Telescope* data suggest, the heated intergalactic gas was inhibited from gravitationally collapsing to form new generations of stars in some small galaxies. The lowest-mass galaxies could not even hold onto their heated gas, and it escaped back into intergalactic space. The radiation causing the ionization of helium came from quasars, which were very abundant during this epoch. Astronomers postulate that quasars are the luminous centers of active galaxies. Active galaxies have extremely energetic central regions, which are probably caused by the presence of a centrally located, supermassive black hole that is feeding and growing by accreting nearby matter. (Black holes are discussed in chapter 6.)

Astrophysicists suggest that the universe was a very active, unruly place during the helium reionization epoch. The black-hole powered cores of active galaxies, called quasars, pumped out enough ultraviolet light to reionize the primordial helium. Specifically, young compact galaxies frequently collided, and these collisions often fed the supermassive black holes in their cores with an abundance of infalling matter. Prior to crossing the black hole's event horizon, the extremely energetic accreting matter would emit large quantities of intense *ultraviolet radiation.* The intense ultraviolet radiation left the colliding galaxies and heated the helium found throughout intergalactic space. The intergalactic helium went from 18,000°F (10,000°C) to about 40,000°F (22,200°C). As the universe expanded, the hot reionized helium gas slowly cooled, and the compact, dwarf galaxies could once again resume their normal star-birthing and assembly activities. Astronomers speculate that without helium reionization, many more dwarf galaxies would have formed—a circumstance significantly altering the universe's currently observed galactic population and structure.

This diagram traces the evolution of the universe from the big bang to the present. Emphasis is placed on two early watershed epochs: hydrogen reionization and helium reionization. *(NASA, ESA, and A. Field [STScI])*

(continued from page 91)

ancient fireball became transparent and continued to cool from a hot 5,400 R (3,000 K) down to a frigid 4.905 R (2.725 K)—the currently measured value of the cosmic microwave background (CMB).

There were not any stars shining when the universe became transparent some 380,000 years after the big bang. Therefore, scientists refer to the ensuing period as the Dark Ages. During the cosmic Dark Ages, density inhomogeneities allowed the attractive force of gravity to form great clouds of hydrogen and helium. Because these clouds also experienced local density inhomogeneities, gravitational attraction gradually formed stars and then slowly gathered groups of these stars into galaxies. Astronomers currently estimate that about 400,000 million years after the big bang conditions were right for a rapid rate of star formation and the cosmic Dark Ages became illuminated by the radiant emissions of millions of young stars.

As the most widely recognized astronomical objects, stars represent the basic building blocks of galaxies. The age, distribution, and composition of the stars in a galaxy trace the history, dynamics, and evolution of that galaxy. Stars are responsible for the manufacture and distribution of chemical elements heavier than hydrogen and helium, such as carbon (C), nitrogen (N), and oxygen (O). Astronomers have discovered that the characteristics of stars are intimately tied to the characteristics of the planetary systems that may form about them. The study of the birth, life, and death of stars is central to the field of astronomy.

Over time, gravitational attraction condensed portions of the giant clouds of primordial (big bang) hydrogen and helium into individual new stars. The very high temperatures encountered in the cores of the more massive of these stars supported the manufacture of heavier nuclei up to and including iron through a process called nucleosynthesis. Elements beyond iron were formed in a more spectacular fashion. Neutron-capture processes deep inside highly evolved, massive stars and subsequent supernova explosions at the end of their relatively short lifetimes synthesized all the elements beyond iron.

Astrophysicists explain that the slow neutron capture process (or s-process) produced heavy nuclei up to and including bismuth-209—the most massive, naturally occurring nonradioactive nucleus. The flood of neutrons that accompanies a supernova explosion is responsible for the rapid neutron capture process (or r-process). It is these rapid neutron-capture reactions that form the radioactive nuclei of the heaviest elements found in nature, such as thorium-232 and uranium-238. The violently explosive force of the supernova also hurls stellar-minted elements throughout interstellar space in a dramatic shower of stardust.

The expelled stardust eventually combines with interstellar gas. This elementally enriched interstellar gas is then available to create a new generation of stars and, for many of these next generation stars, a family of companion planets. About 4.6 billion years ago, humans' solar system (including planet Earth) formed from just such an elementally enriched, giant cloud of hydrogen and helium gas. All things animate and inanimate on Earth are the natural by-products of these ancient astrophysical processes.

SOME PHYSICAL CHARACTERISTICS OF STARS

A star forms when a giant cloud of mostly hydrogen gas, perhaps light-years across, begins to contract under its own gravity. Over millions of years, this clump of mostly hydrogen and helium gas eventually collects into a giant ball that is hundreds of thousands of times more massive than Earth. As the giant gas ball continues to contract under its own gravitational influence, an enormous pressure arises in its interior. Consistent with the laws of physics, the increase in pressure at the center of the protostar is accompanied by an increase in temperature. Then, when the center reaches a minimum temperature of about 18 million rankines (10 million kelvins), the hydrogen nuclei in the center of the contracting gas ball move fast enough so that when they collide these light (low-mass) atomic nuclei can undergo fusion. (As a point of reference, the central temperature of the Sun's core is approximately 27 million R [15 million K].) This is the very moment when a new star is born. The process of nuclear fusion releases a great amount of energy at the center of a young star.

Once thermonuclear burning begins in the star's core, the internal energy release counteracts the continued contraction of stellar mass by gravity. The ball of gas becomes stable—as the inward pull of gravity exactly balances the outward radiant pressure from thermonuclear fusion reactions in the core. Ultimately, the energy released in fusion flows upward to the star's outer surface and the new star "shines."

Size definitely matters in stellar astronomy. Stars come in a variety of sizes, ranging from about one-tenth to sixty (or more) times the mass of humans' parent star, the Sun. It was not until the mid-1930s that astronomers and astrophysicists began to recognize how the process of nuclear fusion takes place in the interiors of all normal stars and fuels their enormous radiant energy outputs. The German-American physicist and Nobel laureate Hans Albrecht Bethe (1906–2005) was the pioneering scientist

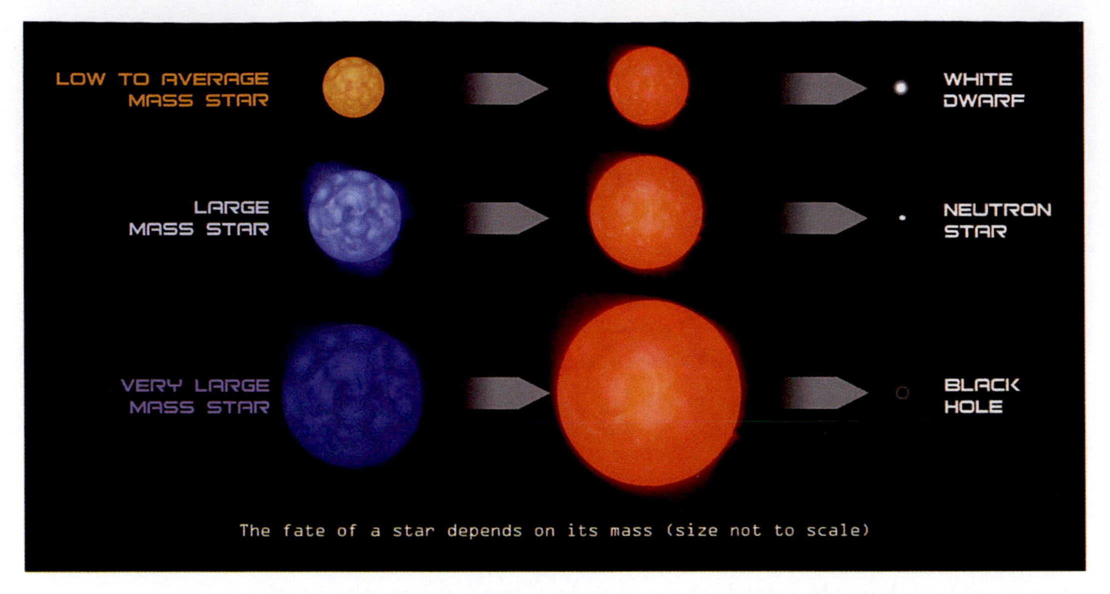

A star's mass determines its ultimate fate (not drawn to scale). *(NASA/CXC)*

who accurately described the physical processes by which stars generate their vast quantities of energy through the nuclear fusion of hydrogen into helium.

A star the size of the Sun requires about 50 million years to reach maturity as it goes from the beginning of the interstellar cloud collapse to main sequence adulthood. Scientists define a main sequence star as one in the prime of its life that shines with a constant luminosity by steadily converting hydrogen to helium through thermonuclear fusion in its core. Stars the mass of the Sun will stay in this mature (main sequence) phase for approximately 10 billion years. Main sequence stars are powered by the continuous conversion of hydrogen into helium deep within their interiors. The outflow of energy from the central regions of the star provides the pressure necessary to keep the star from collapsing under its own mass, as well as the energy by which it shines. A star's luminosity (L) is the rate at which it emits energy per unit time. The Sun has a luminosity (L_{Sun}) of 3.8×10^{26} *watts.*

Main sequence stars span a wide range of luminosities and colors, and astronomers classify them according to those characteristics. The smallest stars, known as red dwarfs, may contain as little as 10 percent of the mass of the Sun and emit only 0.01 percent as much energy, glowing feebly at temperatures between 5,400 R (3,000 K) and 7,200 R (4,000 K). Despite

HUMANS' PARENT STAR—THE SUN

About eight light-minutes away from Earth, the Sun is the nearest star. It has a diameter of 0.86 million miles (1.39 million km) and a mass of 4.38×10^{30} pounds (1.99×10^{30} kg). The Sun is the massive, luminous celestial object about which all other bodies in this solar system revolve. It provides the light and warmth upon which (almost) all terrestrial life depends.

Hydrogen powers the observable universe. The Sun and all other main sequence stars obtain their energy by converting hydrogen into helium through thermonuclear reactions in their intensely hot cores. ***(NASA/ESA)***

The Sun's gravitational field determines the movement of the planets and other celestial bodies (such as comets). Astronomers classify the Sun as a main sequence star of spectral type G2V. Like all main sequence stars, the Sun derives its radiant energy output from thermonuclear fusion reactions involving the conversion of hydrogen to helium and heavier nuclei. Photons associated with these exothermic (energy-releasing) fusion reactions diffuse outward from the Sun's core until they reach the convective envelope. Another by-product of the thermonuclear fusion reactions is a flux of neutrinos that freely escape from the Sun.

At the center of the Sun is the core, where energy is released in thermonuclear reactions. Surrounding the core are concentric shells, which form the radiative zone, the convective envelope (which occurs at approximately 0.8 of the Sun's radius), the photosphere (the layer from which visible radiation emerges), the chromosphere, and, finally, the corona (the Sun's outer atmosphere). Energy is transported outward through the convective envelope by convective (mixing) motions that are organized into cells. The Sun's lower or inner atmosphere, called the photosphere, is the region from which energy is radiated directly into space. Solar radiation approximates a Planck distribution (*blackbody* source) with an effective temperature of 10,440 R (5,800 K).

(continues)

(continued)

The chromosphere, which extends for a few thousand miles (km) above the photosphere, has a maximum temperature of approximately 18,000 R (10,000 K). The corona, which extends several solar radii above the chromosphere, has temperatures of more than 1.8 million R (1 million K). These regions emit *electromagnetic radiation* (EM) in the ultraviolet (UV), extreme ultraviolet (EUV), and X-ray portions of the spectrum. This shorter-wavelength EMR, although representing a relatively small portion of the Sun's total energy output, still plays a dominant role in forming planetary ionospheres and in photochemistry reactions occurring in planetary atmospheres.

Since the Sun's outer atmosphere is heated, it expands into the surrounding interplanetary medium. This continuous outflow of plasma is called the solar wind. It consists of protons, electrons, and alpha particles as well as small quantities of heavier ions. Typical particle velocities in the solar wind fall between 186 and 250 miles per second (300 and 400 km/s), but these velocities may get as high as 620 miles per second (1,000 km/s).

Although the total energy output of the Sun is remarkably steady, its surface displays many types of irregularities. These include sunspots, faculae, plages (bright areas), filaments, prominences, and flares. All are believed ultimately to be the result of interactions between ionized gases in the solar atmosphere and the Sun's magnetic field. Most solar activity follows the sunspot cycle. The number of sunspots varies, with a period of about 11 years. However, this approximately 11-year sunspot cycle is only one aspect of a more general 22-year solar cycle that corresponds to a reversal of the polarity patterns of the Sun's magnetic field.

Sunspots were originally observed by the Italian astronomer Galileo Galilei in 1610. They are less bright than the adjacent portions of the Sun's surface because they are not as hot. A typical sunspot temperature might be 8,100 R (4,500 K) compared to the photosphere's temperature of about 10,440 R (5,800K). Sunspots appear to be made up of gases bubbling up from the Sun's interior. A small sunspot may be about the size of Earth, while larger ones could hold several hundred or even thousands of Earth-size planets. Extra-bright solar regions, called plages, often overlie sunspots. The number and size of sunspots appear to rise and fall through a fundamental 11-year cycle (or in an overall 22-year cycle if polarity reversals in the Sun's magnetic field are considered). The greatest number occurs in years when the Sun's magnetic field is the most severely twisted (called sunspot maximum). Solar physicists think that sunspot migration causes the Sun's magnetic field to reverse its direction. It then takes another 22 years for the Sun's magnetic field to return to its original configuration.

This artist's rendering shows a red dwarf star undergoing a powerful eruption, called a stellar flare. A potentially endangered (hypothetical) exoplanet appears in the foreground. In January 2011, astronomers reported that while observing 215,000 red dwarfs for a seven-day period with the *Hubble Space Telescope* they detected 100 stellar flares. Flares are sudden eruptions of heated plasma that occur when powerful magnetic field lines in a star's atmosphere reconnect, releasing vast amounts of energy. *(NASA, ESA, and G. Bacon [STScI])*

their diminutive nature, red dwarfs are by far the most numerous stars in the universe and have life spans of tens of billions of years.

On the other hand, the most massive stars, sometimes called hypergiants, may be 100 or more times more massive than the Sun and have surface temperatures of more than 54,000 R (30,000 K). Hypergiants emit hundreds of thousands of times more energy than the Sun, but have lifetimes of only a few million years. Although extreme stars such as these are believed to have been common in the early universe, today they are very rare. For example, the entire Milky Way galaxy contains only a handful of hypergiants.

Starting in the 1940s, other scientists began using the term *nucleosynthesis* to describe the complex process of how different-size stars

HANS ALBRECHT BETHE

The German-American physicist and Nobel laureate Hans Albrecht Bethe (1906–2005) proposed the mechanisms by which stars generate their vast quantities of energy through the nuclear fusion of hydrogen into helium. In 1938, he worked out the sequence of nuclear fusion reactions, called the carbon cycle, which dominates energy liberation in stars more massive than the Sun. Later, he recognized the proton-proton reaction as the nuclear fusion process for stars up to the size of the Sun. For this astrophysical work, he received the 1967 Nobel Prize in physics.

create different elements through nuclear fusion reactions and various alpha particle (helium nucleus) capture reactions. All stars on the main sequence use thermonuclear reactions to convert hydrogen into helium, liberating energy in the process. But the initial mass of a star determines not only how long it lives (as a main sequence star) but also how it dies.

Astrophysicists and astronomers consider stars at about the mass of the Sun or less as average- to small-mass stars, which are often collectively referred to as dwarf stars. The Sun, for example, is sometimes referred to as a yellow dwarf star. Astronomers call stars with masses of about one-fifth to one-tenth that of the Sun red dwarf stars. The production of low-mass elements in stars within this mass range is similar. Small- to average-mass stars also share the same fate at the end of their relatively long lives. At birth, small stars begin their stellar life by fusing hydrogen into helium in their cores. This process generally continues for billions of years until there is no longer enough hydrogen in a particular stellar core to fuse into helium. Once hydrogen burning stops, so does the release of the thermonuclear energy that produced the outward radiant pressure, which counteracted the relentless inward attraction of gravity.

At this point in its life, the small star begins to collapse inward. Gravitational contraction causes an increase in temperature and pressure. As a consequence, any hydrogen remaining in the star's middle layers soon becomes hot enough to undergo thermonuclear fusion into helium in a shell around the dying star's core. The release of fusion energy in this shell enlarges the star's outer layers, causing the star to expand far beyond its previous dimensions. This expansion process cools the outer layers of the star, transforming them from brilliant white hot or bright yellow in color to a shade of dull, glowing red. Quite understandably, astronomers call a star at this point in its life cycle a red giant.

Gravitational attraction continues to make the small- to average-mass star collapse, until the pressure in its core reaches a temperature of about 180 million R (100 million K). This very high temperature is sufficient to allow the thermonuclear fusion of helium into carbon. The fusion of helium into carbon now releases enough energy to prevent further gravitational collapse—at least until the helium runs out. This stepwise process continues until oxygen is fused. When there is no more material to fuse at the progressively increasing high temperature conditions within the collapsing core, gravity again exerts its relentless attractive influence on

This 2007 *Hubble Space Telescope* image shows the colorful demise of a sunlike star. Astronomers call the planetary nebula in the image NGC 2440. The white dwarf at the center of NGC 2440 is one of the hottest known, with a surface temperature of about 400,000 R (222,000 K). The planetary nebula's chaotic structure suggests that the dying star shed its mass episodically. *(NASA, ESA, and K. Noll [STSci])*

matter. This time, however, the heat released during gravitational collapse causes the outer layers of the small star to blow off, creating an expanding symmetrical cloud of material that astronomers call a planetary nebula. This expanding cloud may contain up to 10 percent of the dying star's mass. The explosive blow-off process is very important because it disperses into space some of the low-mass elements (such as carbon) that were created within the star.

The final collapse that causes the small star to eject a planetary nebula also liberates thermal energy. But this time the energy release is not enough to fuse other elements. So the remaining core material continues to collapse, until all the atoms are crushed together and only the repulsive force between the electrons counteracts gravity's relentless pull. Astrophysicists refer to this type of very condensed matter as degenerate matter and the gravity-resisting quantum phenomenon as electron degeneracy.

TYPE IA SUPERNOVAS

If the white dwarf star is a member of a binary star system, its intense gravity might pull some gas away from the outer regions of the companion star. When this happens, the intense gravity of the white dwarf causes the inflowing new gas to rapidly reach very high temperatures and a sudden explosion occurs. Astronomers call this event a nova. The nova explosion can make a white dwarf appear up to 10,000 times brighter for a short period of time. Thermonuclear fusion reactions that take place during the nova explosion also create new elements, such as carbon, oxygen, nitrogen, and neon, which are dispersed into space.

In some very rare cases, a white dwarf might undergo a gigantic explosion that astrophysicists call a Type 1a supernova. This happens when a white dwarf that is part of a binary star system pulls too much matter from its stellar companion. Suddenly, the compact "retired" star can no longer support the additional mass and even the repulsive pressure of the electrons in its crushed atoms can no longer prevent further gravitational collapse. This new wave of gravitational collapse heats the helium and carbon nuclei in a white dwarf and causes them to fuse into nickel, cobalt, and iron. However, the thermonuclear burning now occurs so fast that the white dwarf completely explodes. Following a Type Ia supernova, nothing is left behind. All the chemical elements created (by nucleosynthesis) during the lifetime of this small- to average-mass

The final compact object is a degenerate star, called a white dwarf star. The white dwarf star represents the final phase in the evolution of most low-mass stars, including the Sun.

White dwarfs are roughly the size of the Earth but contain most of the mass of a sunlike star. Quantum mechanics (especially the Pauli exclusion principle) provided astrophysicists with the answer they needed to explain why white dwarfs do not continue to collapse under the influence of self-gravitation. In a degenerate star, as the density increases to extreme conditions, the atoms quite literally are squeezed together, causing the orbiting electrons to get closer together. As the orbits of the electrons become reduced, their velocities increase, approaching the speed of light. The electrons reach a point at which their orbits can be squeezed together no further, electron degeneracy pressure kicks in, and further gravitational collapse is halted. It is the electron degeneracy

star are now scattered into space as a result of the spectacular Type Ia supernova detonation.

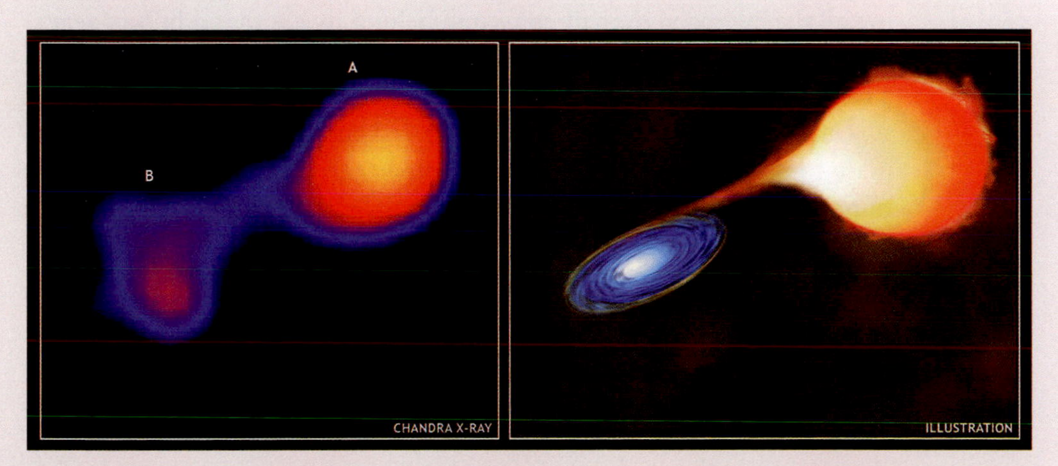

The left side of this composite illustration is a NASA *Chandra X-ray Observatory* image that shows Mira A (upper right), a highly evolved red giant star and Mira B (lower left), a white dwarf. The right side of the composite image is an artist's rendering of the Mira binary star system. Mira A is rapidly losing gas from its upper atmosphere by means of a stellar wind, while Mira B exerts a gravitational tug that creates a gaseous bridge between the two stars. Gas from Mira A accumulates in the high-speed accretion disk around Mira B that produces X-rays. *(X-ray image: NASA/CXC/SAO/M. Karovska et al. Artist's rendering: NASA/CXC/M. Weiss)*

pressure that keeps the cores of dying sunlike stars from completely collapsing.

The more massive the core, the denser the white dwarf that is formed. Thus, the smaller a white dwarf is in diameter, the larger it is in mass. These paradoxical stars are very common—humans' parent star, the Sun, will become a white dwarf billions of years from now. White dwarfs are intrinsically very faint because they are so small. Lacking a source of energy production, they will fade into oblivion as they gradually cool. This fate awaits only those stars with a mass up to about 1.4 times the mass of the Sun. Above that mass, electron degeneracy cannot support the core against further gravitational collapse. Such stars suffer a different, more exotic fate, as described in the next section.

THE FATE OF MASSIVE STARS

Astronomers regard large stars as those having more than about three to five times the mass of the Sun, or more. These stars begin their lives in much the same way as small stars—by fusing hydrogen into helium. However, because of their size, large stars burn faster and hotter, generally fusing all the hydrogen in their cores into helium in less than 1 billion years. Once the hydrogen in the large star's core is fused into helium, it becomes a red supergiant—a stellar object similar to the red giant star, only larger.

Unlike a red giant, the much larger red supergiant star has enough mass to produce much higher core temperatures as a result of gravitational contraction. A red supergiant fuses helium into carbon, carbon and helium into oxygen, and even two carbon nuclei into magnesium. Thus, through a combination of intricate nucleosynthesis reactions the supergiant star forms progressively heavier elements up to and including the element iron. Astrophysicists suggest that the red supergiant has an onionlike structure with different elements being fused at different temperatures in layers around the core. The process of *convection* brings these elements from the star's interior to near its surface, where strong stellar winds then disperse them into space.

Thermonuclear fusion continues in a red supergiant star until the element iron is formed. Iron is the most stable of all the elements. Elements lighter than (below) iron on the periodic table generally emit energy when joined or fused in thermonuclear reactions, while elements heavier than (above) iron on the periodic table emit energy only when their nuclei split or fission (see appendix on page 199).

Scientists suggest that neutron capture is the way elements more massive than iron are created. Neutron capture occurs when a free neutron (one outside an atomic nucleus) collides with an atomic nucleus and sticks. This capture process changes the nature of the compound nucleus, which is often radioactive and undergoes decay, thereby creating a different element with a new atomic number.

This mosaic image is one of the largest ever taken by NASA's *Hubble Space Telescope* of the Crab Nebula in the constellation of Taurus. The image shows a six-light-year-wide expanding remnant of a star's supernova explosion. The orange filaments are the tattered remains of the star and consist mostly of hydrogen. A rapidly spinning neutron star (the crushed, ultra-dense core of the exploded star) is embedded in the center of the nebula and serves as the dynamo powering the nebula's eerie interior bluish glow. Chinese astronomers witnessed this violent event in 1054 C.E. *(NASA, ESA, J. Hester [Arizona State University])*

Slow neutron capture (s-process) reactions that take place deep in the interior of a highly evolved star gradually produce heavier elements up to bismuth (Bi). However, the heaviest chemical elements (such as thorium [Th] and uranium [U]) are formed by rapid neutron capture reactions when a literal flood of neutrons are released during a supernova explosion.

This is what happens when a large star goes supernova. The red supergiant eventually produces the element iron in its intensely hot core. Because of nuclear stability phenomena, the element iron is the last chemical element formed that can be formed by nucleosynthesis. When nuclear reactions begin to fill the core of a red supergiant star with iron, the overall thermonuclear energy release in the large star's interior begins to decrease. Because of this decline, the star no longer has the internal radiant pressure to resist the attractive force of gravity. Thus, the red supergiant begins to collapse. Suddenly, this gravitational collapse causes the core temperature to rise to more than 180 billion R (100 billion K), smashing the electrons and protons in each iron atom together to form neutrons. The force of gravity now draws this massive collection of neutrons incredibly close together. For about a second, the neutrons fall very fast toward the center of the star. Then they smash into one another and suddenly stop. This sudden stop causes the neutrons to recoil violently, and an explosive shock wave travels outward from the highly compressed core. As this shock wave travels from the core, it heats and accelerates the outer layers of material of the red supergiant star.

The traveling shock wave causes the majority of the large star's mass to be blown off into space. Astrophysicists refer to this enormous explosion as a Type II supernova. The stellar end product of a Type II supernova is either a neutron star or a black hole (see chapter 6).

A supernova will often release (for a brief period) enough energy to outshine an entire galaxy. Since supernova explosions scatter star-manufactured elemental materials far out into space, they represent the primary way heavier chemical elements disperse in the universe.

Both Type la and Type II supernovas release an enormous amount of energy. For a period of days to weeks, a supernova may outshine an entire galaxy. Likewise, all the naturally occurring elements and a rich array of subatomic particles are produced in these explosions. On average, a supernova explosion occurs about once every hundred years in the typical galaxy. About 25 to 50 supernovas are discovered each year in other galaxies, but most are too far away to be seen without a telescope.

This artist's concept shows jets of material shooting out from the neutron star in the binary system 4U 0614+091. Using data from NASA's *Spitzer Space Telescope* in 2006, astronomers discovered these remarkable jets, which are streaming into space at nearly the speed of light. The 4U 0614+091 system contains two dead stars. The object shown in the upper left is a white dwarf—the remnant core of a sunlike star. The neutron star (lower right at the center of the accretion disk) is the remnant of a much more massive star that exploded in a Type II supernova. The more dense and massive neutron star is drawing material from its companion to fuel and feed its energetic jets. *(NASA/JPL-Caltech/R. Hunt [SSC])*

The dust and debris left behind by supernovas eventually blend with the surrounding interstellar gas and dust, enriching it with the heavy elements and chemical compounds produced during stellar death. Eventually, those materials are recycled, providing the building blocks for a new generation of stars and accompanying planetary systems.

If the collapsing stellar core at the center of a supernova contains between about 1.4 and 3 solar masses, the collapse continues until electrons and protons combine to form neutrons, producing a neutron star. The phenomenon of neutron degeneracy pressure prevents the remnant core from further collapsing under the relentless influence of gravity.

Neutron stars are incredibly dense and have a density similar to the density of an atomic nucleus. Because this type of exotic star contains so much mass packed into such a small volume, the pull of gravity at its surface is immense. Like the white dwarf stars, if a neutron star forms in a multiple star system, it can accrete gas by stripping it off any nearby companions.

Neutron stars also have powerful magnetic fields that can accelerate atomic particles around its magnetic poles, producing powerful beams of radiation. Those beams sweep around like massive searchlight beams as the star rotates. If such a beam is oriented so that it periodically points toward Earth, astronomers observe it as regular pulses of radiation that occur whenever the magnetic pole sweeps past the line of sight. In this case, scientists call the neutron star a pulsar.

If the collapsed stellar core is larger than three solar masses, the large dying star collapses completely to form an incredibly exotic cosmic object known as a black hole. The black hole is an infinitely dense object, whose gravity is so strong that nothing, not even light, can escape its immediate proximity. Fortunately for science, indirect observations of these massive, mysterious objects are possible because the gravitational field of a black hole is so powerful that any nearby material (frequently often the outer layers of a companion star) is caught up and dragged in. As matter spirals toward the event horizon of a black hole, it forms an accretion disk that is heated to enormous temperatures, emitting copious quantities of X-rays and gamma rays that hint at the presence of a massive, invisible dark cosmic companion. Black holes are discussed in more detail in chapter 6.

EXOPLANETS SPARK THE SECOND COPERNICAN REVOLUTION

In the mid-16th century, the Polish astronomer and church official Nicholas Copernicus (1473–1543) published a book suggesting that Earth, like the other known planets (Mercury, Venus, Mars, Jupiter, and Saturn), orbited around the Sun. His bold hypothesis helped unravel almost 2,000 years of geocentric cosmology; stimulated the Scientific Revolution; and caused much technical, social, and political turmoil.

Up until the mid-1990s, many astronomers were uncomfortable postulating that other stars had planetary companions. While planet formation is a physical logical corollary to star formation, the astronomers simply did not have any observational evidence to validate the important hypothesis. Then, through an exciting series of telescopic discoveries, they

QUARK STARS

In the early 1960s, the American physicist Murray Gell-Mann introduced quark theory to help describe the behavior of hadrons (nuclear particles such as neutrons and protons) within the atomic nucleus. Based upon the current standard model of elementary particles, the neutron consists of three confined quarks: one up quark and two down quarks. The quark gluon plasma (QGP) is an extremely dense state of matter in which all the quarks are unconfined and

(continues)

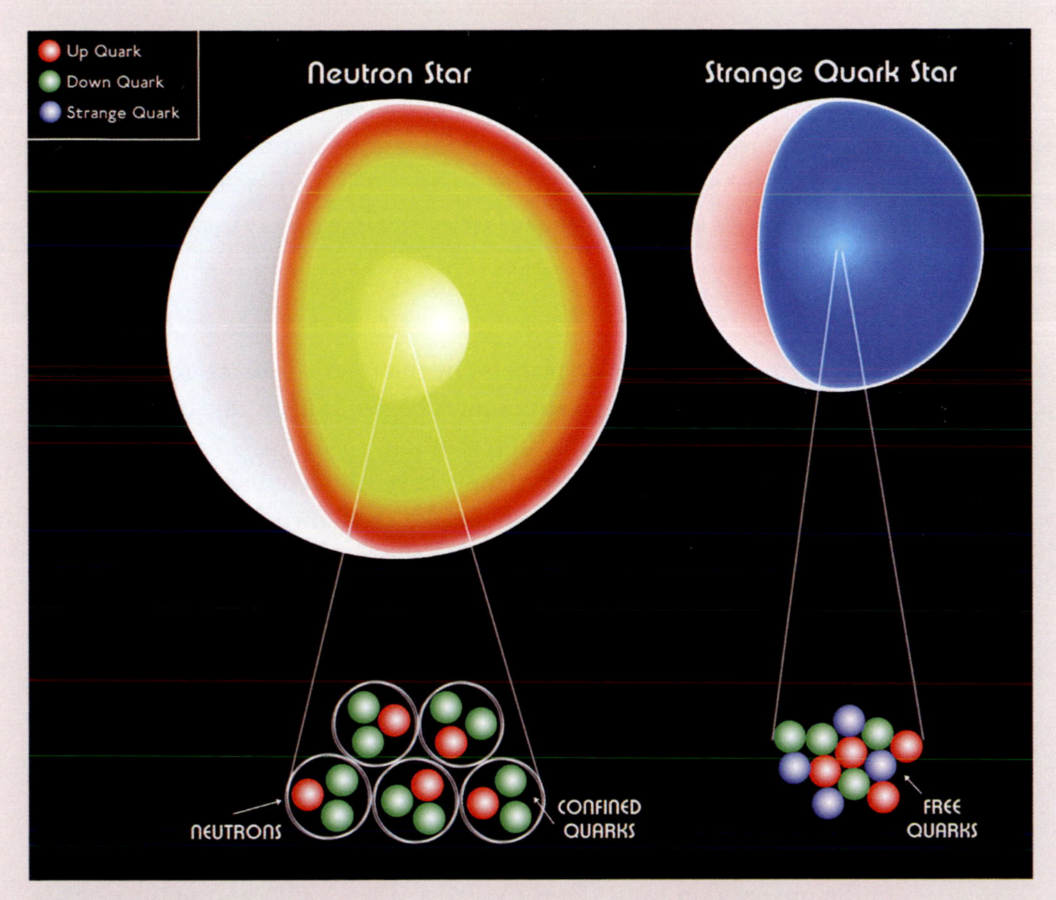

This illustration compares a neutron star (about 12 miles [19.3 km] in diameter) to a strange, or quark star (about seven miles [11.3 km] in diameter). In a neutron star (shown on left), the quarks are confined inside the neutrons; in a quark star (shown on right), the quarks are free or unconfined in a quark-gluon plasma. Since unconfined quarks take up less space, a quark star has a smaller diameter but higher density than a neutron star. ***(NASA/CXC/M. Weiss)***

(continued)

move about in a very high temperature soup. Particle physicists suggest that such extreme conditions existed in the first 10 microseconds or so after the big bang. Once the early universe became a bit cooler, protons and neutrons formed, confining the free quarks.

Theoretical physicists propose that some neutron stars in certain binary star systems might accrete enough gas from the companion stars to add sufficient pressure on the dead star's surface to squeeze the confined quarks in the interior, making them unconfined or free quarks. This hypothesized ultracompact cosmic object, called the strange or quark star, would contain a very hot, dense QGP, consisting of free (unconfined) up, down, and strange quarks. The hypothesized quark star represents an intermediate step between the neutron star and the black hole. Some recent astronomical observations support the quark star hypothesis, but additional observations are needed before scientists reach a consensus on the validity of the quark star hypothesis.

began detecting extrasolar planets. In their active search for Earthlike planets around alien suns, astronomers began to liken this period of discovery to the second Copernican revolution.

Throughout human history, people have wondered about the possible existence of other worlds, like or unlike their own. The earliest scientific demonstrations of Copernicus's heliocentric hypothesis validated that other worlds did indeed orbit about the Sun. The ensuing centuries of planetary astronomy and more recently decades of space exploration revealed that all these companion worlds were quite different from Earth, and also very different from one another. As scientists began to understand that the relatively fixed, very distant lights (stars) in the sky were other suns, and that numerous galaxies consisted of billions of stars, it seemed a near certainty that other planets should exist around stars.

However, this tantalizing hypothesis could not be proven through scientific observation until the early 1990s. Then, radio and optical astronomical observations detected small changes in stellar emission, which revealed the presence of first a few examples and now many examples of planetary systems around other stars. Scientists call these planets exoplanets to distinguish them from the planetary worlds in humans' solar system.

This is an artist's rendering of the pulsar planet system discovered by the Polish astronomer Aleksander Wolszczan in 1992. The three planets, the first ever found outside humans' solar system, orbit a rapidly rotating neutron star, known as pulsar PSR B1257+12. *(NASA/JPL-Caltech)*

Although all the technical details are still not entirely understood, scientists recognize that stars form from spinning protostellar disks of interstellar gas and dust. Earth and other planets of this solar system are thought to have developed from the remains of the Sun's ancient protostellar disk. Scientists have uncovered no plausible reason why the same process of star and planet formation that took place in humans' solar system cannot and is not taking place throughout the Milky Way galaxy and in other galaxies.

The detection of the first exoplanet system revealed a planetary world utterly unlike this solar system. The Polish astronomer Aleksander

Wolszczan (1946–) was working with the Canadian astronomer Dale Frail in the early 1990s at the Areceibo Observatory in Puerto Rico. They were using the giant radio telescope to make precise measurements of the pulses from the millisecond pulsar PSR1257+12. The data they collected indicated that this 1-billion (10^9)-year-old pulsar had at least two planetary companions. They reported this important discovery in the January 9, 1992, issue of *Nature.* As an interesting note of historic coincidence, Wolszczan had earned his doctoral degree at the Copernicus University in Toruń, Poland. Today, astronomers know that this pulsar has three planetary companions. Pulsars are neutron stars that spin rapidly and pulse with radiation. As shown in the figure on page 111, the farthest two planets from the pulsar (closest in the view) are about the size of Earth. Radiation from charged particles traveling along the pulsar's magnetic field lines would most likely rain down on the planets, causing their skies to glow with auroras. One such aurora is depicted at the bottom of the picture.

Subsequent discoveries continue to reveal the existence of many other exoplanet systems—each quite dissimilar from humans' solar system. In some alien solar systems, planets as massive as Jupiter (or larger) orbit so close to their parent star that they are heated to high temperatures, sweeping their upper atmospheres into interplanetary space. Astronomers called such unusual planets hot Jupiters. In other alien solar systems, planets follow elongated orbits. This is in stark contrast to the nearly circular planetary orbits found in humans' solar system. Exoplanet-hunting activities are just beginning, and no astronomer can now predict whether Earthlike planets are common or very rare and special. The discovery of Earthlike planets at suitable distances from their parent stars would invigorate speculation about the possibility of life elsewhere in the universe (see also chapter 10).

CHAPTER 6

The Dark Side of the Universe

"Eliminate all other factors, and the one which remains must be the truth."

—*Sherlock Holmes, fictional detective*
From Sir Arthur Conan Doyle's
"The Sign of the Four" (1890)

This chapter discusses three of the most puzzling and mysterious phenomena that govern the behavior of the universe: black holes, dark matter, and dark energy. The current scientific understanding of each baffling phenomenon is sketchy at best. One of the paramount goals of scientists in this century is to develop a successful theoretical framework that correctly explains how each functions. In their own mysterious ways, black holes, dark matter, and dark energy appear to have influenced the evolution of the universe and now control the ultimate fate of all matter and energy.

Scientists devised the concepts of dark matter and dark energy to help them explain, in the simplest ways possible, the astronomical motions they were observing in the 20th century. As hypothesized, dark matter is the dominant, but invisible, mass whose gravitational influence governs the observed motions of galaxies and galaxy clusters. Similarly, scientists introduced the concept of dark energy to identify the mysterious antigravitational phenomenon that was causing the universe to expand at an accelerated rate.

BLACK HOLES—GRAVITY'S ULTIMATE TRIUMPH OVER MATTER

One of the most fascinating objects in astronomy is the black hole. As scientists now understand, a black hole is a massive, gravitationally collapsed object from which nothing—light, matter, or any other kind of signal—can escape.

Astrophysicists conjecture that a stellar black hole is the natural end product when a giant star dies and collapses. If the core (or central region) of the dying star has three or more solar masses left after exhausting its nuclear fuel, then no known force can prevent the core from forming the extremely deep (essentially infinite) gravitational warp in spacetime, called a black hole. (One solar mass contains approximately 4.39×10^{30} lbm [1.99×10^{30} kg].) As with the formation of a white dwarf or a neutron star (discussed in chapter 5), the collapsing giant star's density and gravity increase with gravitational contraction. Because of the larger mass involved in this case, the force of gravity of the collapsing star becomes too strong for even neutron degeneracy pressure to resist and an incredibly dense point mass, or singularity, forms.

Physicists define *singularity* as a point of zero radius and infinite density—in other words, it is a point where spacetime is infinitely distorted. Basically, a black hole is a singularity surrounded by an event region in which the gravity is so strong that absolutely nothing can escape. Due to the limits of contemporary science, no one now understands what goes on inside a black hole. Since no light or any other radiation can escape, scientists do not have any directly observable signals from the inside of a black hole. That is why the black hole is quite literally black.

Nevertheless, physicists try to study the black hole by speculating about its theoretical properties and then by looking for perturbations in the observable universe that provide telltale hints that a massive, invisible object matching such theoretical properties is possibly causing such perturbations. For example, on Earth, a pattern of ripples on the surface of an otherwise quiet but murky pond could indicate that large fish is swimming just below the surface. Similarly, astrophysicists look for detectable ripples in the observable portions of the universe to support theoretical predictions about the behavior of black holes.

PRIMITIVE BLACK HOLES

In 2010, scientists using data from NASA's *Spitzer Space Telescope* discovered two of the earliest supermassive black holes known. Dating back about 13 billion years ago, these very primitive cosmic monsters are called J0005-0006 and J0303-0019. They are among the most distant cosmic objects yet observed by astronomers.

These ancient black holes appear to be in the very earliest stages of formation. However, unlike other more recently formed black holes, this ancient duo lacks dust. As the accompanying illustration suggests, gas swirls around

(continues)

This artist's rendering depicts one of the most primitive supermassive black holes (central black dot) known. The cosmic monster lies at the core of a young, star-rich galaxy. Astronomers used data from NASA's *Spitzer Space Telescope* to discover two such primitive objects. They are called J0005-0006 and J0303-0019 and date back about 13 billion years. ***(NASA/JPL-Caltech/R. Hurt [SSC])***

(continued)

a black hole in the region called the accretion disk. Usually the accretion disk is surrounded by a dark, doughnutlike structure, called a dust torus. But for a very primitive black hole early in its development, the anticipated dust torus is absent and astronomers only observe a gas disk.

The proposed reason for this is that not enough time had elapsed in the early universe (some 700 million years after the big bang) to allow molecules to clump together into dust particles. Thus, black holes forming during this period started out with accretion disks that lacked dust. Then, as such primitive black holes grew, they devoured more and more mass. It was only during these adolescent feeding periods that the ancient black holes began attracting dusty rings to accompany their gaseous accretion disks.

The very concept of a mysterious black hole exerts a strong pull on both scientific and popular imaginations. Data from spacebased observatories, such as NASA's *Chandra X-ray Observatory* (CXO), have moved black holes from the purely theoretical realm to a dominant position in observational astrophysics. Scientists have accumulated strong evidence that black holes not only exist, but very large ones, called supermassive black holes (containing millions or billions of solar masses), may function like cosmic monsters that lurk at the centers of every large galaxy.

The basic idea of a black hole is more than two centuries old. The first person to publish a "black hole" paper was John Michell (1724–93), a British geologist, amateur astronomer, and clergyman. His 1784 paper suggested the possibility that light (then erroneously believed to consist of tiny particles of matter subject to influence by Newton's law of gravitation) might not be able to escape from a sufficiently massive star. Michell was a competent astronomer who successfully investigated binary star system populations. He further suggested in this 18th-century paper that although no "particles of light" could escape from such a massive object, astronomers might still infer its existence by observing the gravitational influence it exerted on nearby celestial objects.

The French mathematician and astronomer Marquis Pierre-Simon de Laplace (1749–1827) introduced a similar concept in the late 1790s, when he also applied Newton's law of gravitation to a celestial body so massive that the force of gravity prevented any light particles from escaping. Neither Laplace nor Michell used the term *black hole* to describe their postulated, very massive heavenly body. The term *black hole* did not enter the lexicon of science until the American physicist John Archibald Wheeler (1911–2008) introduced it in 1967. Both of these 18th-century black hole speculations were on the right track but suffered from incomplete and inadequate physics.

The needed breakthrough in physics took place a little more than a century later when Albert Einstein introduced general relativity. Einstein replaced the Newtonian concept of gravity as a force with the novel notion that gravity was associated with the distortion of the spacetime. Einstein proposed that the more massive an object, the greater its ability to distort the local spacetime continuum. Spacetime (also written space-time) is the synthesis of the three spatial dimensions and a fourth dimension, time. Most people find it difficult to simultaneously consider these four dimensions. Einstein's genius allowed him to apply the concept of spacetime in both his special and general relativity theories.

Shortly after Einstein introduced general relativity in 1915, the German astronomer Karl Schwarzschild (1873–1916) discovered that Einstein's general relativity equations led to the postulated existence of a dense object into which other objects could fall but out of which no objects could ever escape. In 1916, Schwarzschild wrote the fundamental equations that describe a black hole. He also calculated the size of the event horizon, or boundary of no return, for this incredibly dense and massive object. The dimension of the event horizon now bears the name Schwarzschild radius in his honor.

The notion of an event horizon implies that no information about events taking place inside this distance can ever reach outside observers. However, the event horizon is not a physical surface. Rather it represents the start of a special region that is disconnected from normal space and time. Although scientists cannot see beyond the event horizon into a black hole, they believe that time stops there.

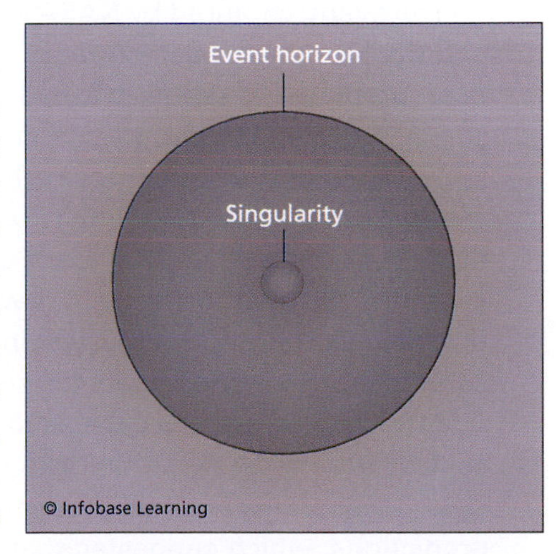

This is a basic diagram of a Schwarzschild black hole (not drawn to scale). The singularity is much smaller than the size of an atomic nucleus; the event horizon marks the boundary between the normal universe and the unusual world within the black hole. At the event horizon, the escape velocity equals the speed of light. For a 10 solar-mass black hole, the radius of the event horizon is 18.6 miles (30 km). *(Based on DOE artwork)*

Inside the event horizon, the escape speed exceeds the speed of light. Outside the event horizon, escape is possible. It is important to remember that the event horizon is not a material surface. It is the mathematical def-

SUPERMASSIVE BLACK HOLE RIPS UNLUCKY STAR APART

In 2004, astronomers reported for the first time that X-ray data from orbiting spacecraft allowed them to witness a supermassive black hole ripping apart a star and consuming a portion of it. The artist's rendering (upper illustration in the composite image) shows that a yellow star in the center of galaxy RXJ1242-11 was bumped off course by another star and thrown into the path of a supermassive black hole. The enormous gravity of the black hole then stretched the star until it was torn apart and partially devoured.

The event, captured by NASA's *Chandra X-ray Observatory* (lower left image) and the European Space Agency's *XMM-Newton X-ray Observatory,* had long been predicted in theory but had not previously been observed. Astronomers suggest that following a close encounter with another star the doomed star (about the mass of the Sun) came too close to the supermassive black hole in the center of the galaxy RXJ1242-11. As the unlucky star neared the supermassive black hole (estimated total mass of 100 million solar masses), it was severely stretched by tidal forces until it was torn apart. This observational discovery provided astronomers important information about how large black holes might continue to grow and impact the surrounding stars and interstellar gas.

After reviewing the available data, astronomers estimated that about 1 percent of the mass of the unlucky star was ultimately consumed by the black hole. This relatively small amount of mass is consistent with their theoretical predictions, which suggested that the momentum and energy of the accretion process would cause most of the doomed star's gas to be flung away from the black hole. Gas from the star that was devoured by the black hole was initially heated to millions of degrees R (K), while spinning around in the black hole's accretion disk prior to disappearing across the event horizon. The heated gases emitted X-rays that were detected as a powerful outburst from the center of galaxy RXJ1242-11. Supporting optical data (lower right image) from the European Southern Observatory (ESO) show a pair of galaxies, including the suspected location (white circle displayed in lower right image) of the supermassive black hole in RXJ1242-11.

inition of the point of no return, the point where all communication is lost with the outside world. Inside the event horizon, the laws of physics, as scientists currently understand them, do not apply. Once anything crosses

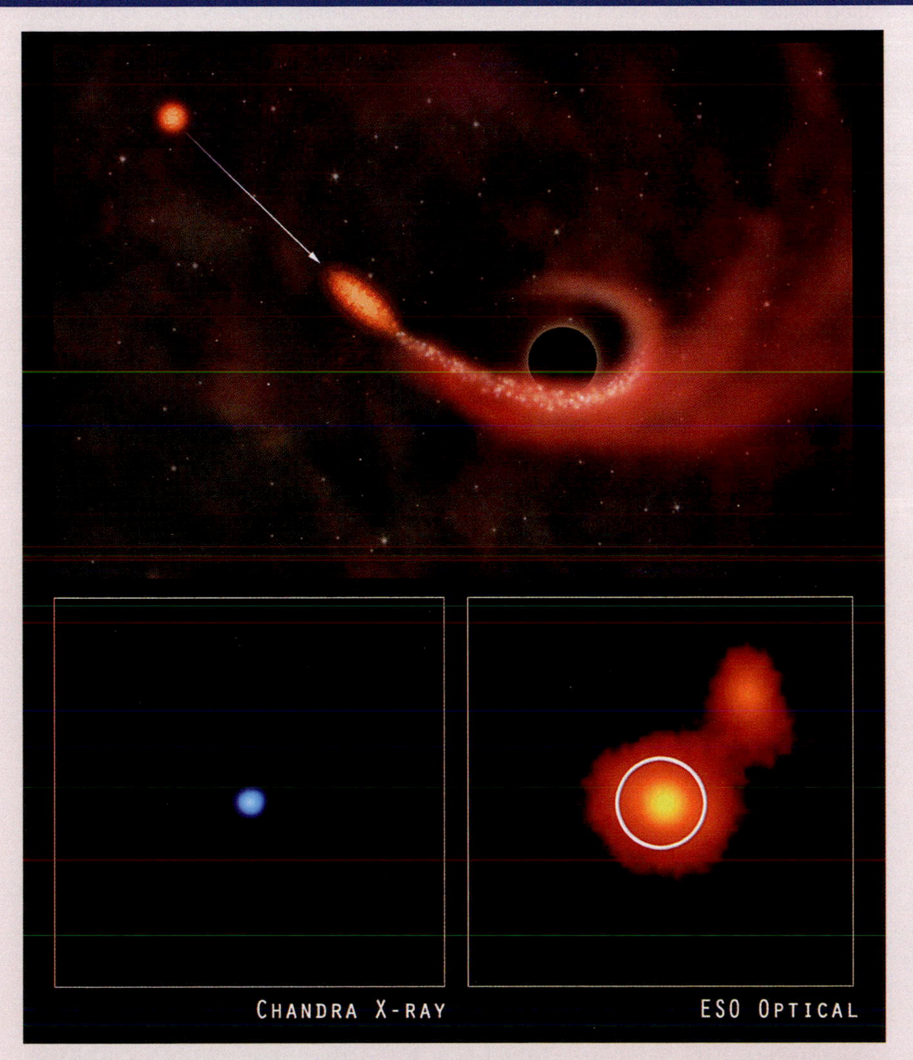

The artist's rendering (upper image) shows a supermassive black hole destroying an unlucky star in galaxy RXJ1242-11. X-ray (lower left) and optical (lower right) data provided astronomers the first strong evidence of a supermassive black hole ripping apart a star and consuming a portion of it. ***(Artist's rendering: NASA/CXC/M. Weiss; X-ray data: NASA/CXC/MPE/S. Komossa, et al.; optical data: ESO/MPE/S. Komossa)***

this boundary, it will disappear into an infinitesimally small point—the singularity, previously mentioned. Scientists cannot observe, measure, or test a singularity since it, too, is a mathematical definition. The little scientists know about black holes comes from looking at the effects they have on their surroundings outside the event horizon. As more powerful space-based observatories, such as NASA's planned *James Webb Space Telescope,* study the universe across all portions of the electromagnetic spectrum in the 21st century, scientists will be able to construct better and better theoretical models of the black hole.

The mass of the black hole determines the extent of its event horizon. The more massive the black hole, the greater the extent of its event horizon. The event horizon for a hypothetical Schwarzschild black hole containing 10 solar masses is just 18.6 mi (30 km) from the singularity. This size black hole is an example of a stellar black hole. Astronomers have identified dozens of stellar black hole candidates (containing between three to 20 solar masses or more) by observing how some stars appear to wobble as if nearby, yet invisible, massive companions were pulling on them. X-ray binary systems offer another way to search for candidate black holes.

In comparison to stellar black holes, the radius of the event horizon for a supermassive black hole consisting of 1 billion (10^9) solar masses is about 1.86 billion miles (3 billion km), which is comparable to the distance of Uranus from the Sun. Astronomers suspect that such massive objects exist at the centers of many galaxies because they provide one of the few logical explanations for the strange and energetic events now being observed there. No one knows for certain exactly how such supermassive black holes formed. One hypothesis is that over billions of years, relatively small stellar black holes (formed by supernovas) began devouring neighboring stars in the star-rich centers of large galaxies and eventually became supermassive black holes.

Once matter crosses the event horizon and falls into a black hole, only three physical properties appear to remain relevant: total mass, net electric charge, and total angular momentum. Recognizing that all black holes must have mass, physicists have proposed four basic stellar black hole models. The Schwarzschild black hole (first postulated in 1916) is a static black hole that has no charge and no angular momentum. The Reissner–Nordström black hole (introduced in 1918) has an electric charge but no angular momentum—that is, it is not spinning. In 1963, the New Zealander mathematician Roy Patrick Kerr (1934–) applied general relativity to describe the properties of a rapidly rotating, but uncharged,

This artist's rendering shows a nonrotating black hole (on the left) and a rotating black hole (on the right). The most detailed studies of stellar black holes to date indicate that not all black holes spin at the same rate. *(NASA/CXC/M. Weiss)*

black hole. Astrophysicists think this model is the most likely "real-world" black hole because the massive stars that formed them would have been rotating. One postulated feature of a rotating Kerr black hole is its ringlike structure (the ring singularity) that might give rise to two separate event horizons. Some astrophysicists have even suggested it might become possible (at least in theory) to travel through the second event horizon and emerge into a new universe or possibly a different part of this universe. The final black hole model has both charge and angular momentum. Called the Kerr-Newman black hole, this theoretical model appeared in 1965. Since astrophysicists currently think that rotating black holes are unlikely to have a significant electric charge, the uncharged Kerr black hole remains the more favored "real-world" candidate model for the stellar black hole.

Astrophysicists think they have found indirect, but reliable, ways of detecting stellar black holes. The best currently available techniques depend upon candidate black holes being members of binary star systems. Unlike the Sun, many stars (more than 50 percent) in the Milky Way galaxy are members of a binary system. If one of the stars in a particular binary system has become a black hole, although invisible, it would betray its existence by the gravitational effects it produces upon the observable companion star. Once beyond event horizon, the black hole's gravitational influence is the same as that exerted by other objects (of equivalent mass) in the normal universe. So a black hole's gravitational effects on its companion would simply follow Newton's universal law of gravitation—

ASTRONOMERS INVESTIGATE X-RAY JETS OF BLACK HOLE

In 2002, astronomers reported that they successfully tracked the life cycle of X-ray jets from a stellar black hole in the Milky Way galaxy. Using data from NASA's *Chandra X-ray Observatory* (left side of the accompanying composite image), they watched as the jets evolved and traveled at near light speed for several years before slowing down. NASA's *Rossi X-ray Timing Explorer* had initially detected an outburst of X-rays in 1998 from the double-star system called XTE J1550-564. (An artist's rendering of this X-ray binary system appears on right side of composite image.) The ejection of jets from stellar and supermassive black holes appears to be a rather common occurrence throughout the observable universe. As the jets plow through nearby interstellar gas, the resistance of the gas slows them down. The observations of XTE J1550-564 represent the first time astronomers were able to witness such jets in the act of slowing down. Scientists examine relatively nearby stellar black holes in this galaxy in the hope of studying similar processes occurring in distant quasars and active galactic nuclei.

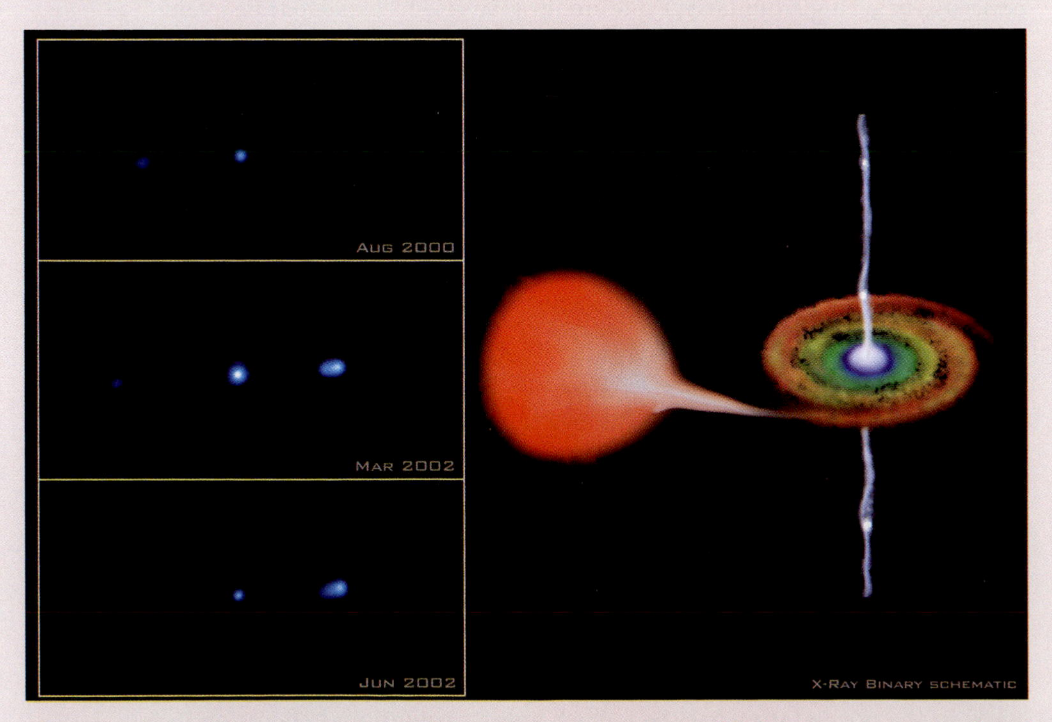

Chandra X-ray Observatory **data (left) and artist's rendering (right) show jets associated with the stellar black hole in X-ray binary system XTE J1550-564.** ***(NASA/CXC/M. Weiss)***

that is, the mutual gravitational attraction of the two celestial objects is directly proportional to their masses and inversely proportional to the square of the distance between them.

Astrophysicists now speculate that a substantial part of the energy of matter spiraling into a black hole is converted by collision, compression, and heating into X-rays and gamma rays that display certain spectral characteristics. X-ray and gamma radiations emanate from the material as it is pulled toward the black hole. But once the captured material crosses the black hole's event horizon, this energetic telltale radiation cannot escape.

Suspected black holes in binary star systems exhibit this type of prominent material capture effect. Astronomers have discovered several black hole candidates, using space-based astronomical observatories (such as the *Chandra X-ray Observatory*). One very promising candidate is called Cygnus X-1, an invisible object in the constellation Cygnus (the Swan). The notation Cygnus X-1 means that it is the first X-ray source discovered in Cygnus. X-rays from the invisible object have characteristics like those expected from materials spiraling toward a black hole. This material is apparently being pulled from the black hole's binary companion, a large star of about 30 solar masses. Based upon the suspected black hole's gravitational effects on its visible stellar companion, the black hole's mass has been estimated to be about six solar masses. In time, the giant visible companion might itself collapse into a neutron star or a black hole; or else it might be completely devoured by its black hole companion. This form of stellar cannibalism would also significantly enlarge the existing black hole's event horizon.

In 1963, the Dutch-American astronomer Maarten Schmidt (1929–   ) was analyzing observations of a star named 3C 273. He had very confusing optical and radio data. What he and his colleagues had discovered was the quasar. Today, astronomers know that the quasar is a type of active galactic nucleus (AGN) in the heart of a normal galaxy. AGN galaxies have hyperactive cores and are much brighter than normal galaxies. The AGNs emit energy equivalent to converting the entire mass of the Sun into pure energy every few years. Energy is emitted in all regions of the electromagnetic spectrum, from low-energy radio waves all the way to much higher-energy X-rays and gamma rays. Furthermore, the energy output of these AGN can vary on short time scales (hours and days), suggesting that the source is very compact. Stars by themselves, powered by nuclear fusion, cannot generate such levels of energy. Even the impressive supernovae explosions are insufficient. So astrophysicists puzzled over what physical processes could produce the *power* of more than 100 Milky Way galaxies and do it within a region of space only a few light-years across?

To further compound this cosmic mystery, some of these AGN galaxies have extraordinary jets of material rushing out of their cores that stretch far into space—up to 100 to 1,000 times the diameter of the galaxy. After considering all types of energetic processes, including the simultaneous explosions of thousands of supernovae, most astrophysicists now think that supermassive black holes represent the most plausible answer. Although nothing emerges from a black hole, matter falling into one can release tremendous quantities of energy just before it crosses the event horizon. The region just outside the event horizon will glow in X-rays and gamma rays—the most energetic forms of electromagnetic radiation.

Matter captured by the gravity of a black hole will eventually settle into a disk around the black hole. Scientists call the inner region of swirling, superheated material the accretion disk. Stellar black holes can have accretion disks if they have a nearby companion star. Material from the companion will be drawn into orbit around the black hole, thereby forming the accretion disk. The diameter of the accretion disk depends on the mass of the black hole. The more massive the black hole, the larger the accretion disk. The accretion disk of a stellar black hole will stretch out only a few hundred or thousand kilometers from the center. However, the accretion disk of a supermassive black hole is much bigger and becomes solar system–size.

The most spectacular accretion disks exist in active galaxies suspected of containing supermassive black holes. The *Hubble Space Telescope (HST)* has provided astronomers strong evidence for this assumption. The galaxy NGC 4261 is an elliptical galaxy, whose core contains an unexpected large disk of dust and gas. Astronomers think that a black hole may lurk within the central region of this galaxy. Radio observations of this galactic core have also revealed jets of material ejected from the center of this disk. This provides corroborating evidence for the existence of a very large black hole.

Exactly what creates and controls the flow of matter out of these jets is still not clearly understood. It is almost as if black holes are messy eaters, consuming matter but spewing out leftovers. Mostly likely, the jets have something to do with the rotation and/or the magnetic fields of the black hole. Whatever the actual cause, most astronomers believe that only black holes are capable of producing such spectacular and outrageous behavior.

This false-color infrared image from NASA's *Spitzer Space Telescope* shows hundreds of thousands of stars crowded into the swirling core of the Milky Way galaxy. Old and cool stars appear blue, while dust features lit up by hot, massive stars appear in a reddish hue. The brightest white spot in the middle is the very center of the galaxy, which also marks the site of a supermassive black hole. *(NASA/JPL-Caltech/S. Stolovy [Spitzer Science Center/Caltech])*

Scientists using the *Hubble Space Telescope (HST)* have also discovered a 3,700 light-year-diameter dust disk encircling a suspected 300 million solar mass black hole in the center of the elliptical galaxy NGC 7052, located in the constellation Vulpecula about 191 million light-years from Earth. This disk is thought to be the remnant of an ancient galaxy collision, and it will be swallowed up by the giant black hole in several billion years. *Hubble Space Telescope* measurements have shown that the disk rotates rapidly at 96 miles per second (155 km/s) at a distance of 186 light-years (ly) from the center. The speed of rotation provides scientists a direct measure of the gravitational forces acting on the gas due to the presence of a suspected supermassive black hole. Though 300 million times the mass of the Sun, this suspected black hole is still only about 0.05 percent of the total mass of the NGC 7052 galaxy. The bright spot in the center of the giant dust disk is the combined light of stars that have been crowded around the black hole by its strong gravitational pull. This stellar concentration appears to match the theoretical astrophysical models that link stellar density to a central black hole's mass.

In the 1990s, X-ray data from the German-American *Roentgen Satellite (ROSAT)* and the Japanese-American *Advanced Satellite for Cosmology and Astrophysics (ASCA),* suggested that a mid-mass black hole might exist in the galaxy M82. This observation was confirmed in September 2000, when astronomers compared high-resolution *Chandra X-ray Observatory* images with optical, radio, and infrared maps of the region. Scientists now think such black holes must be the results of black hole mergers since they are far too massive to have been formed from the death of a single star. Sometimes called the "missing link" black holes, these medium-size black holes fill the gap in the observed black hole masses between stellar and supermassive. The M82 missing link is not in the absolute center of the galaxy, where all supermassive black holes are suspected of residing; but it is comparatively close to it.

The detailed, multispectral study of the environmental influence exerted by a suspected black hole on the nearby universe just beyond its event horizon, including the dragging of the spacetime continuum, is an exciting area of contemporary space-based astronomy. Unanticipated discoveries made in the next few decades could easily change the trajectory of contemporary physics—influencing the trajectory of human civilization in the process.

DARK MATTER

Scientists currently define dark matter as the material in the universe that cannot be observed directly because it emits very little or no electromagnetic radiation but whose gravitational effects can be measured and quantified. Dark matter was originally called missing mass and, as implied by that earlier name, was discovered through its gravitational effects.

While investigating the cluster of galaxies in the constellation Coma Berenices (Berenice's Hair) in 1933, the Swiss astrophysicist Fritz Zwicky (1898–1974) noticed that the velocities of the individual galaxies in this cluster were so high that they should have escaped from one another's gravitational attraction long ago. He concluded that the amount of matter actually present in the cluster had to be much greater than what could be accounted for by the visibly observable galaxies. From his observations, Zwicky estimated that the visible matter in the cluster was only about 10 percent of the mass actually needed to gravitationally bind the galaxies

ZWICKY AND THE INITIAL HUNT FOR MISSING MASS

Shortly after Edwin Hubble introduced the concept of an expanding universe in the late 1920s, the Swiss astrophysicist Fritz Zwicky pioneered the early search for the missing mass of the universe. Today, scientists refer to Zwicky's missing mass as dark matter. In 1934, while collaborating with the German-American astronomer Walter Baade (1893–1960), Zwicky remained at the frontiers of astrophysics by postulating the creation of a neutron star as a result of a supernova explosion. Not only was he a truly visionary physicist, he was also a scientist with an extremely eccentric and often caustic personality. In the 1960s, Zwicky published an extensive catalogue of galaxies and clusters of galaxies.

together. A galaxy cluster is an accumulation of galaxies (from 10 to hundreds or even a few thousand members) that lie within a few million light-years of one another and are bound by gravitation. Without the presence of Zwicky's hypothesized missing matter, the galaxies in a cluster would have drifted apart and escaped from one another's gravitational attraction a long time ago.

Starting in the 1960s, the American astronomer Vera (Cooper) Rubin (1928–) observed dark matter's distinctive gravitational signature in the stars of the Andromeda galaxy (M31). Specifically, she noticed that the stars in the outer portions of this majestic spiral galaxy were traveling at unexpectedly high speeds. Their unusual rotational velocities suggested that something else was providing a much stronger gravitational influence than did the quantity of mass implied by the galaxy's observable (that is, visible) stars. Zwicky, Rubin, and other scientists used the rotational speeds of individual galaxies within a cluster of galaxies (as obtained from their Doppler shifts) to provide observational evidence that most of the mass of the universe might actually be in the form of invisible material, now generally called dark matter.

The second direct observational evidence of the existence of dark matter came from careful radio astronomy-supported studies of the rotation rate of individual galaxies, including the Milky Way galaxy. From their rotational behavior, astronomers discovered that most galaxies appear to be surrounded by a giant cloud (or galactic halo)—containing matter capable of exerting gravitational influence but not emitting observable radiation. These studies also indicated that the great majority of a galaxy's mass lays in this very large halo, which is perhaps 10 times the diameter of the visible galaxy. The Milky Way galaxy is a large spiral galaxy that contains about 100 billion stars. Observations indicate humans' home gal-

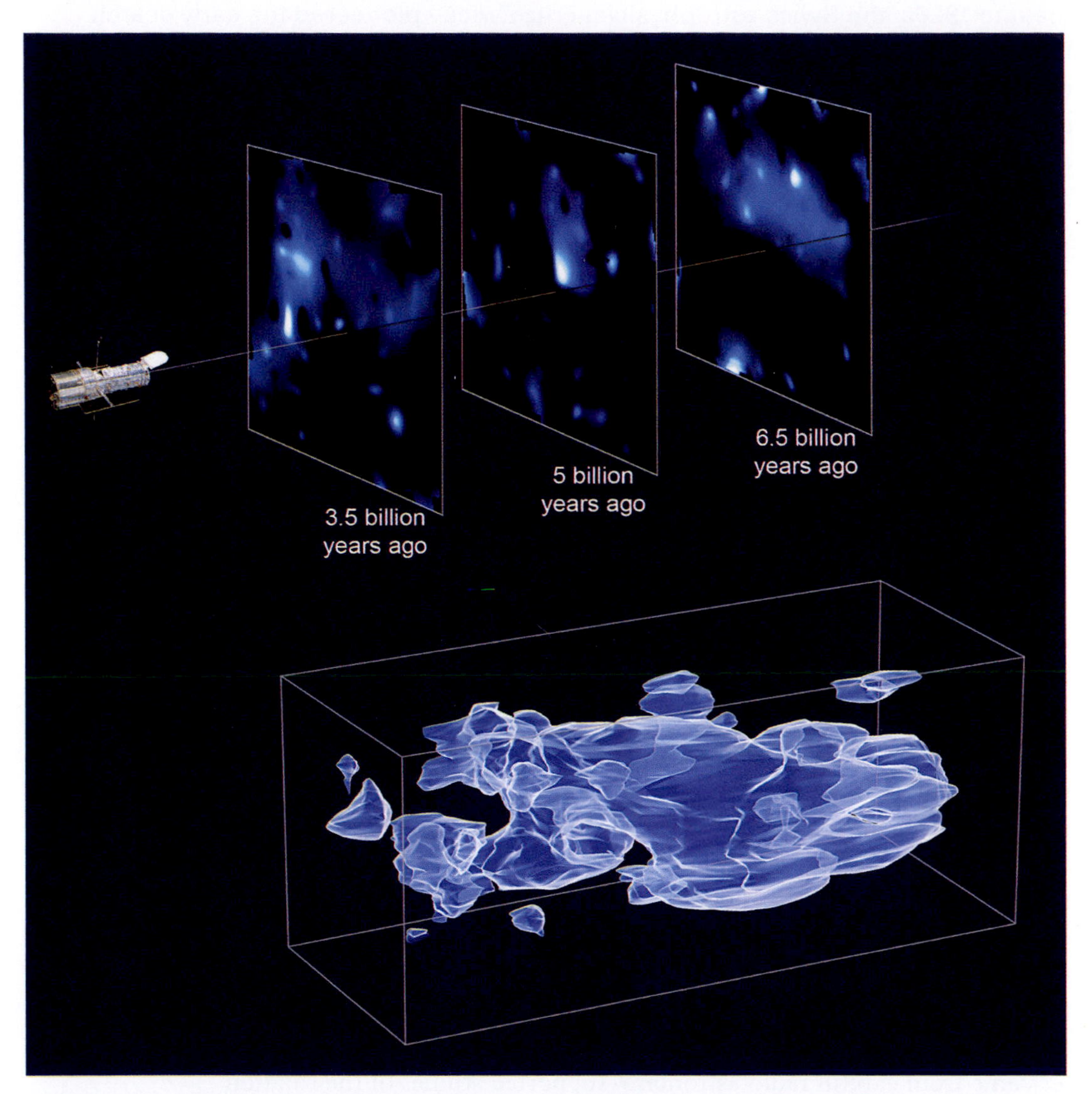

Hubble Space Telescope (HST) data allowed scientists to map the three-dimensional distribution of dark matter in the universe as a function of time. They used a method called weak gravitational lensing to observe how dark matter deflected the light of distant galaxies. The top (sliced) image shows how dark matter evolved from 6.5 billion to 3.5 billion years ago. The bottom image shows dark matter clumping as it collapses under gravity. *(NASA/ESA/CalTech)*

axy is surrounded by a dark matter halo, which probably extends out to about 750,000 light-years. The mass of this dark matter halo appears to be about 10 times greater than the estimated mass of all the visible stars in the Milky Way. However, the material composition of this influential dark matter halo still remains a great astronomical mystery.

It should come as no surprise that there is considerable disagreement within the scientific community as to what this dark matter really is. Two general schools of scientific thought have emerged—one advocating MACHOs (or baryonic matter) and one advocating WIMPs (or nonbaryonic matter).

The first group assumes dark matter consists of MACHOs or *ma*ssive *c*ompact *h*alo *o*bject*s*—essentially ordinary matter that astronomers have simply not yet detected. This unobserved but ordinary matter is composed of heavy particles (baryons), such as neutrons and protons. The brown dwarf is one candidate MACHO that could significantly contribute to resolving the missing mass problem.

The brown dwarf is a substellar (almost a star) celestial body that has the material composition of a star but that contains too little mass to permit its core to initiate thermonuclear fusion. In 1995, astronomers detected the first brown dwarf candidate object—a tiny companion orbiting the small red dwarf star Gliese 229. Also very difficult to detect are low-mass white dwarfs that represent another dark matter candidate. Using sophisticated observational techniques that take advantage of *gravitational lensing* (discussed in chapter 9), astronomers have collected data suggesting that low-mass white dwarfs could make up about half of the dark matter in the universe. Finally, black holes, especially relatively low-mass primordial black holes that formed soon after the big bang, represent another MACHO candidate.

The second group of scientists speculates that dark matter consists primarily of exotic particles that they collectively refer to as WIMPS (*w*eakly *i*nteracting but *m*assive *p*article*s*). These exotic particles represent a hypothetical form of matter called nonbaryonic matter—matter that does not contain baryons (protons or neutrons). Some of the WIMP candidates proposed by scientists include: neutrinos, axions, and neutralinos.

Physicists postulated the existence of neutrinos in the early 20th century to help them fill intellectual gaps in nuclear particle physics. Later in the century, neutrinos were experimentally detected. At that time, they were assumed to be weakly interacting, massless particles. In 1998,

(continues on page 132)

HUNTING DARK MATTER WITH THE *HUBBLE SPACE TELESCOPE*

Scientists currently think that the most common substance in the universe is dark matter—exceeding the total amount of ordinary (or baryonic) matter by a factor of five or more. Yet, this very strange, nonbaryonic substance does not shine nor does it reflect light. Since scientists cannot see dark matter, they generally cannot use with the same instruments or techniques that they use to study ordinary matter.

This *Hubble Space Telescope (HST)* composite image shows a ghostly, deep blue "ring" of dark matter superimposed upon a visible *HST* image of galaxy cluster Cl 0024+17. The scientist-generated, deep blue map of the cluster's dark matter distribution is based upon a phenomenon called gravitational lensing. Although invisible, dark matter is inferred as present because its gravitational influence bends the incoming light from more distant galaxies. ***(NASA/ESA/M.J. Lee and H. Ford [Johns Hopkins University])***

A number of scientists have speculated that dark matter is a substance composed of special subatomic particles, or possibly even atoms, which are very much different from those that make up the ordinary matter found throughout the observable universe on planets, in stars, and in galaxies. Despite such interesting speculations, no one really knows what this strange form of matter is. If a person could somehow construct a giant wall out of dark matter and then drive an ordinary-matter automobile into it at high speed, the person would not dent a bumper, crack the vehicle's headlights, or inflate any of vehicle's impact-protection airbags. Incredibly, the driver would not even know this hypothetical high-speed collision with dark matter had taken place. At present, the only credible fact known about dark matter is that it can exert a significant, large-scale gravitational influence.

Scientists wanted to learn what really happens to dark matter when it experiences a collision with ordinary matter. They used data from NASA's *Hubble Space Telescope (HST)* and were able to get a firsthand view of how dark matter behaves during a gigantic collision between two galaxy clusters. The colossal cosmic wreck created a ripple of dark matter—a ripple somewhat analogous to the ripple formed on the surface of a pond when a rock hits the water. The *HST* composite image (shown here) depicts a deep blue, ghostly ring of dark matter in the galaxy cluster Cl 0024+17. The ringlike structure of the (scientist-generated) deep blue map corresponds to the cluster's dark matter distribution. To create this composite image, NASA scientists carefully superimposed the shadowy deep blue map on top of a more familiar, visible image of the same galaxy cluster.

The approach taken by modern scientists is somewhat analogous to the way authorities used the invisible man's telltale footsteps in the snow to hunt down the rogue scientist in the 1933 motion picture version of H.G. Wells's famous science fiction story *The Invisible Man* (1897). Modern scientists now use the telltale "footprints" made by dark matter's gravitational influence to hunt their invisible quarry.

The discovery of the ring of dark matter in galaxy cluster Cl 0024+17 represents some of the strongest evidence yet collected testifying to the existence of dark matter. Since the 1930s, astronomers and astrophysicists have suspected the existence of some invisible substance, which serves as the source of the additional gravity needed to hold galaxy clusters together. Scientists estimated that the masses associated with the visible stars in these clusters was simply not enough to provide the gravity forces needed to keep these whirling systems together. Without some invisible source of additional, gravity-providing mass, the laws of physics predict that such galaxy clusters must fly apart because of their motions.

Many improvements with astronomical instruments and observatories have been made since the 1930s. But despite these great improvements, scientists still cannot directly observe dark matter. However, the advances in instrumentation now allow them to infer its existence in galaxy clusters by observing how the gravitational influence of invisible dark matter bends the light of more distant background galaxies. Scientists generated this map of the galaxy cluster's dark matter distribution by making very careful observations of how gravitational lensing affected the light from more distant galaxies.

(continued from page 129)
scientists discovered that one type of neutrino had a very tiny amount of mass. If particle physicists confirm that all neutrinos actually possess tiny, nonzero rest masses, then these ubiquitous but very weakly interacting, elementary particles might account for much of the universe's dark matter.

To help explain the absence of an electrical dipole moment for the neutron, some physicists have postulated another particle called the axion. Axions would have very little mass but might have been produced in copious quantities during the big bang. Neutralinos are members of another set of hypothetical particles, which some physicists have proposed in connection with a theory called supersymmetry (see chapter 9). Neutralinos are envisioned as being much heavier than the proton but still represent the lightest of the postulated electrically neutral supersymmetric particles. Physicists have yet to observe either the axion or the neutralino.

Some scientists suggest that the true nature of dark matter may not require an "all or nothing" characterization. To these physicists, it seems perfectly reasonable that dark matter might exist in several forms, including difficult to detect low-mass stellar and substellar objects (MACHOs) in the inner regions of a dark matter galactic halo, as well as swarms of exotic particles (WIMPs) farther out in the galactic halo. Discovering the true nature of dark matter remains a very important and intriguing challenge for 21st-century cosmologists and astrophysicists.

DARK ENERGY

Dark energy is the generic name now being given by astrophysicists and cosmologists to an unknown cosmic force field thought to be responsible for the recently observed acceleration in the rate of expansion of the universe. The American astronomer Edwin P. Hubble (1889–1953) first proposed the concept of an expanding universe in 1929. He suggested that observations of Doppler-shifted wavelengths of the light from distant galaxies indicated that that these galaxies were receding from Earth with speeds proportional to their distance.

In the late 1990s, while two competing teams of scientists were making systematic surveys of very distant Type Ia (carbon detonation) supernovas, they observed that instead of slowing down (as might be anticipated if gravity was the only significant force at work in cosmological dynamics), the rate of recession (that is, the redshift) of these very distant objects appeared to be actually increasing. It was almost as if some unknown force was neutralizing or canceling the attraction of gravity.

Such startling observations proved controversial and very inconsistent with the then popular gravity-only models of an expanding universe within big bang cosmology. Despite fierce initial resistance within the scientific community, these perplexing observations eventually gained acceptance. Today, carefully analyzed and reviewed supernova data indicate that the rate of expansion of the universe is accelerating—a dramatic conclusion that has tossed modern cosmology into a great turmoil.

Cosmologists do not yet have an acceptable answer as to what could be causing this apparent accelerated expansion. Some scientists revisited the cosmological constant (symbol Λ). Albert Einstein inserted this concept into his original general relativity theory to make his revolutionary theory of gravity describe a static universe—that is, a nonexpanding one, which had neither a beginning nor an end. But after boldly introducing the cosmological constant as representative of some mysterious force associated with empty space capable of balancing or even resisting gravitational attraction, Einstein decided to abandon the idea. Hubble's announcement of an expanding universe provided the intellectual nudge that encouraged Einstein's decision. Afterward, Einstein personally referred to the notion of a cosmological constant as "my greatest failure."

Physicists are now revisiting Einstein's concept and suggesting that there is possibly a vacuum pressure force (a recent name for the cosmological constant). This force appears inherently related to empty space but seems to exert its influence only on a very large scale. Consequently, the mysterious force would have been negligible during the very early stages of the universe following the big bang event; but it would later manifest itself and serve as a major factor in cosmological dynamics. Since such a mysterious force is neither required nor explained by any of the currently known laws of physics, scientists do not yet have a clear physical interpretation of just what such a mysterious (gravity-resisting) force really means.

As mentioned in chapter 2, on March 7, 2008, NASA released the results of a five-year investigation of the oldest light in the universe—the cosmic microwave background (CMB). Based on a careful evaluation of the CMB data collected by a NASA spacecraft called the *Wilkinson Microwave Anistropy Probe (WMAP),* scientists were able to gain incredible insight into the past and present content of the universe. The *WMAP* data revealed that the current contents of the universe include about 5 percent "ordinary" atoms, the building blocks of stars, planets, and people. Dark matter comprises about 23 percent of the universe. Finally, the remaining 72 percent of the universe is composed of dark energy, which acts like a matter-repulsive, antigravity phenomenon. One of the central challenges facing scientists in this century is to develop a comprehensive understanding of the cosmic roles of ordinary matter, dark matter, and dark energy.

Cosmologists understand that the ultimate fate of the universe is far from understood but clearly being influenced by dark energy. If dark energy eventually weakens or reverses its antigravitational behavior, the

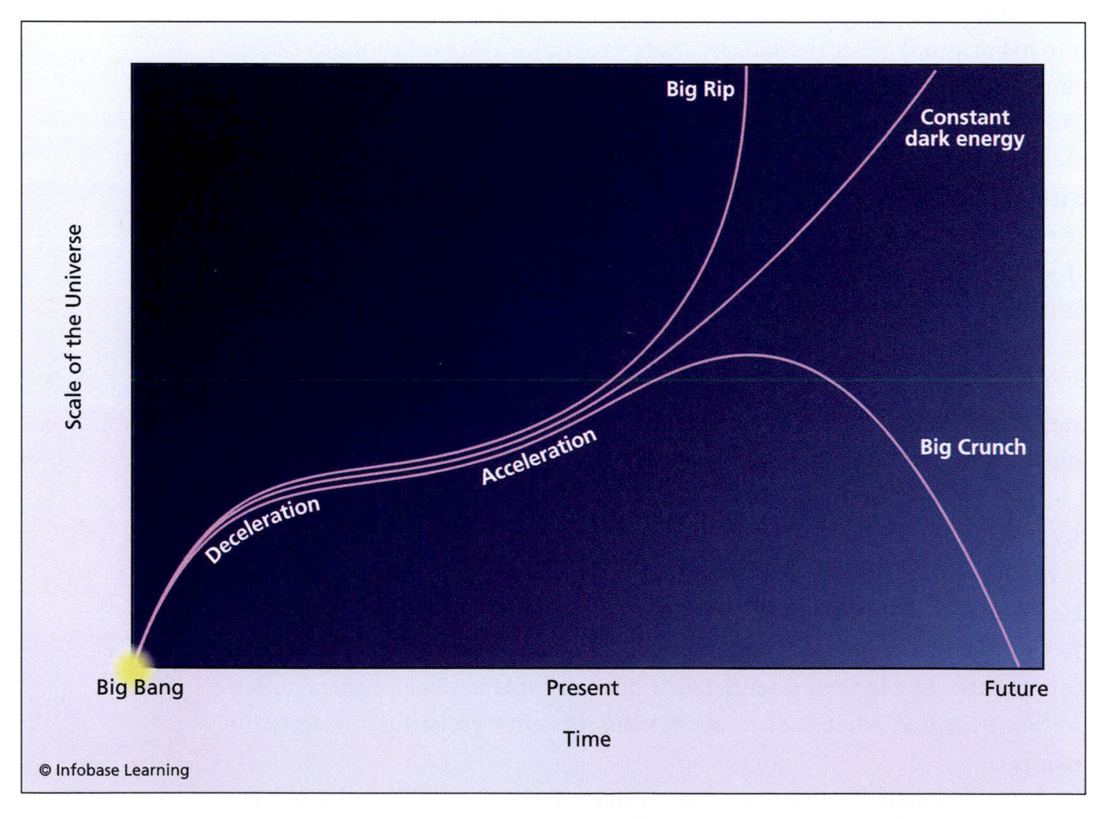

This graph illustrates the currently perceived role of dark energy in the evolution of the universe. Three future scenarios emerge. If dark energy weakens or reverses its antigravitational behavior, the big crunch results; if dark energy grows in strength, the big rip occurs; if dark energy remains constant, the universe will continue to expand forever but in a gradual manner. *(NASA/CXC/M. Weiss)*

universe could experience an overall gravitational pull that causes all matter and energy to collapse back into a singularity. Scientists call this fate the big crunch.

If dark energy continues to accelerate the expansion of the universe, galaxies would fly away from one another. In the final extreme, dark energy would keep gaining strength and cause all matter in the universe to repel all other matter. Individual galaxies would disassemble, individual solar systems would disintegrate, and finally individual atoms would tear themselves apart. Scientists call this fate the big rip.

Cosmologists have also developed a middle-ground scenario. In the constant dark energy scenario, the overall influence of dark energy remains constant, despite the increasing size of the universe and corresponding dilution of mass's gravitational influence. The universe would continue to expand forever but in a slower, less dramatic manner than suggested by the big rip scenario.

Scientists anticipate that an understanding of dark energy will lead to a much deeper understanding of nature and the basic behavior of the universe. They presently rely on two fundamental theories to explain how the universe works: quantum mechanics and general relativity (Einstein's theory of gravity). Quantum mechanics describes the universe on the smallest of scales, while general relativity describes the universe on the largest of scales. Since quantum mechanics and general relativity are not compatible with each other, some scientists suggest that dark energy might serve as the bridge between these two dominant theories.

CHAPTER 7

Very Cold Matter

This chapter discusses very cold matter and how scientists quantified the familiar but elusive concept of cold. Topics include the general perception of cold, the physical properties and applications of cryogenic fluids, superconductivity, and the Bose-Einstein condensate—an unusual, extremely low-temperature state of matter. Thermodynamic implications of low-temperature environments are also addressed.

UNDERSTANDING COLD

If asked to describe cold, most people would say that cold represents the absence of heat or warmth. This definition is obviously quite subjective. Scientists would elaborate a bit further, quantifying how cold a substance or environment actually is by specifying its temperature. Using the well-established laws of thermodynamics, scientists state that the coldest possible condition corresponds to a temperature value of absolute zero.

Four centuries of temperature-related scientific research eventually provided people with amazing new insights into the nature of cold. In the 19th century, remarkable advances in low-temperature engineering gave human beings unprecedented mastery over cold, including such important technical developments as refrigeration, air-conditioning, and liquefied gases, including liquid oxygen and liquid nitrogen. Starting in the 20th century, scientific efforts in low-temperature physics produced more than two dozen Nobel Prizes, including the 2001 award in physics for "the achievement of Bose-Einstein condensation in dilute gases of alkali atoms." By investigating the behavior of matter near absolute zero, physicists began to venture into a very strange quantum mechanical world, where certain extremely cold fluids (called superfluids) defied gravity and certain materials (called superconductors) allowed the flow of electricity without resistance at low temperatures.

Since the mid-19th century, the pursuit of absolute zero represented the holy grail for low-temperature physicists. Scientists define absolute zero as the temperature at which molecular motion ceases and an object has no thermal energy. In 1848, the Scottish physicist Lord Kelvin (William Thomson) (1824–1907) became the first scientist to formally suggest its existence. Based on the well-proven laws of thermodynamics, absolute zero is the lowest possible temperature; by international scientific agreement, it is 0 K (–273.15°C) on the Kelvin temperature scale. As a historic footnote, in 1859, the Scottish engineer and physicist William John Macquorn Rankine (1820–72) introduced an engineering version of Kelvin's absolute temperature scale—a less frequently used scale that carries Rankine's name. Absolute zero on the Rankine temperature scale is 0 R (–459.67°F) and corresponds to absolute zero on the Kelvin temperature scale.

Due primarily to the work of the German physical chemist Walther Hermann Nernst (1864–1941) and the German physicist Max Planck, the third law of thermodynamics emerged in the early 1900s. Scientists state this law as follows: The entropy of any pure substance in thermodynamic equilibrium approaches zero as its temperature approaches absolute zero. For example, a perfect crystal at absolute zero would have perfect order (that is, zero entropy). The third law is important in that it provides a basis for calculating the absolute entropies of substances, either elements or compounds. Scientists then use these data in analyzing chemical reactions.

At absolute zero, atoms and molecules are at their lowest (but finite) energy state. In practice, scientists recognize that it is impossible to achieve absolute zero because the input power requirements to obtain this theoretical temperature limit would approach infinity. However, research scientists have achieved temperatures that lie within a few billionths of a degree above absolute zero. Extremely low-temperature researchers speak in terms of millikelvin (mK) (10^{-3} K), microkelvin (μK) (10^{-6} K), nanokelvin (nK) (10^{-9}), and even picokelvin (pK) (10^{-12}) environments.

For example, a team of NASA-funded scientists reported in 2003 that they created a Bose-Einstein condensate (BEC) at an incredibly frigid temperature of only 450 pK. (BECs are discussed later in the chapter.)

SOME NATURALLY COLD ENVIRONMENTS

On Earth, Antarctica is synonymous with the concept of bitter cold. It is not only the coldest but also the windiest, highest (on average), and driest region on Earth. The Antarctic continent consists of 5.4 million square miles (14 million km^2) of ice, snow, and rock that encircle the planet's South Pole. Some 98 percent of Antarctica is covered by a thick layer of ice, which has an average thickness of about one mile (1.6 km). The elevation of the Antarctic continent ranges from 6,560 feet to 13,120 feet (2,000 m to 4,000 m) with the mountain ranges reaching nearly 16,400 feet (5,000 m). Scientists regard Antarctica as a desert because precipitation along the coastal regions averages only eight inches (20.3 cm) per year, and it is much less in the interior of the continent. The lowest natural temperature ever recorded on Earth's surface was measured by Russian scientists at the Vostok Station in Antarctica. On July 21, 1983, they recorded a record low temperature of –129.3°F (–89.6°C).

A newly born Weddell seal pup in Antarctica. Weddell seals live farther south than any other mammal. They are fat animals that have become naturally adapted to the Antarctic's frigid conditions. *(NOAA)*

Only a few cold-adapted animals and plants survive in Antarctica's inhospitable coastal regions. Except for a few temporary human researchers, Weddell seals are the most southward living mammals. These hardy creatures have coats of fur around their rather plump bodies. Adult Weddell seals are between 880 and 1,320 pounds-mass (400–600 kg); at birth, Weddell pups have a mass of between 55 and 65 lbm (25 and 30 kg). A variety of birds (including several species of penguin) and other types of seals also inhabit the coastal zones of the frozen continent. Antarctica has no indigenous human population, although upward of 5,000 scientists occupy a number of research stations throughout the year.

Among the currently explored solar system objects, Neptune's frigid moon Triton appears to have the coldest surface conditions. It is possible that other, even colder bodies will be discovered later this century when robot spacecraft travel beyond Neptune's orbit and visit various trans-Neptunian objects. For example, NASA's *New Horizons* spacecraft is on schedule to encounter the dwarf planet Pluto in 2015 and then travel beyond into the Kuiper belt. The Kuiper belt occurs in the outer solar system beyond Neptune, out to perhaps 1,000 astronomical units away from the Sun. This region contains millions of icy objects, which range in size from tiny particles to Pluto-size dwarf planets. The Dutch-American astronomer Gerard Peter Kuiper (1905–73) first suggested the existence of this disk-shaped reservoir of icy objects in 1951.

Triton is the largest moon of Neptune. It has a diameter of approximately 1,680 miles (2,700 km) and a surface temperature of only –391°F (–235°C [38 K])—making it the coldest object yet explored in the solar system. Triton was discovered by the British astronomer William Lassell (1799–1880) in 1846, barely a month after the planet Neptune itself was discovered by the German astronomer Johann Gottfried Galle (1812–1910). This large moon is in a unique synchronous retrograde orbit around Neptune at a distance of 220,350 miles (354,760 km). The moon's orbit is characterized by a period of 5.877 days and an inclination of 157.4° (to Neptune's orbit). Triton is the only large moon in the solar system that rotates in a direction opposite to the direction of rotation of its planet, while keeping the same hemisphere toward the planet. Triton has a tenuous atmosphere, consisting of primarily nitrogen with small amounts of methane and traces of carbon monoxide.

The accompanying computer-generated montage shows Neptune (background) as it would appear from a spacecraft approaching Triton. The wind and sublimation-eroded south polar cap of Neptune's largest

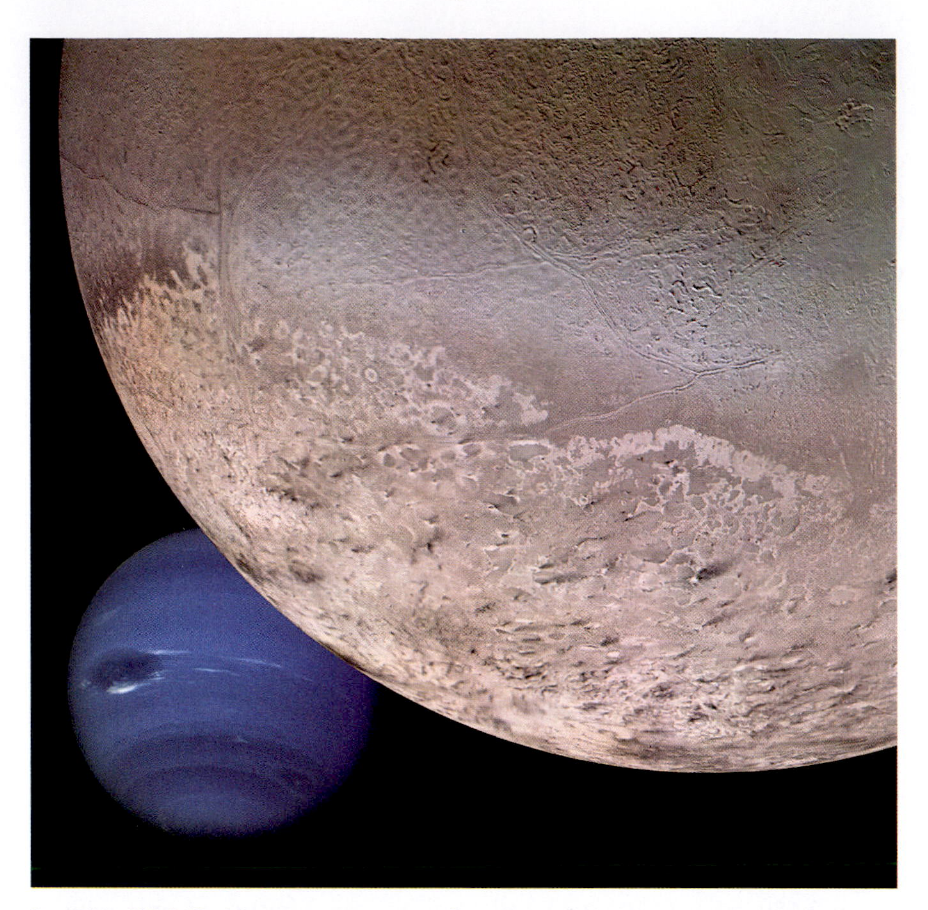

In 1989, NASA's *Voyager 2* spacecraft collected the data upon which this computer-generated montage is based. The figure shows Neptune (background) as it would appear from a spacecraft approaching Triton. The moon's surface temperature is just –391°F (–235°C [38 K]), making it the coldest known surface in the solar system. *(NASA/JPL/USGS)*

moon, Triton (foreground), appears at the bottom of the image. Cryovolcanic terrain is evident at the upper right, while cantaloupe terrain is found at the upper left. Triton's frigid surface is mostly covered by nitrogen frost mixed with traces of condensed methane, carbon dioxide, and carbon monoxide. Triton's very tenuous atmosphere is thick enough to produce wind-deposited streaks of dark and bright materials (of currently unknown origin) in the south polar cap region. At the time of the *Voyager 2* flyby in 1989, Triton's southern polar cap was sublimating, as suggested by the irregular and eroded appearance of the edge of the cap. Scientists synthesized the color in the accompanying montage by combining filtered, high-resolution images taken by the *Voyager 2* spacecraft. The cryovolcanic landscapes were apparently produced when icy-cold liquids (now frozen) erupted from Triton's interior.

The coldest overall temperature found in nature is the temperature of outer space itself. Scientists have carefully measured the residual primal glow of the big bang fireball. Called the cosmic microwave background, this radiation acts like an opaque wall that surrounds the entire observable universe and now has an average blackbody temperature of only –454°F (–270°C [2.7 K]).

However, the story of very cold natural locations does not end here. About 5,000 light-years away from Earth lies the Boomerang Nebula (sometimes called the Bow Tie Nebula). This amazing cosmic object has

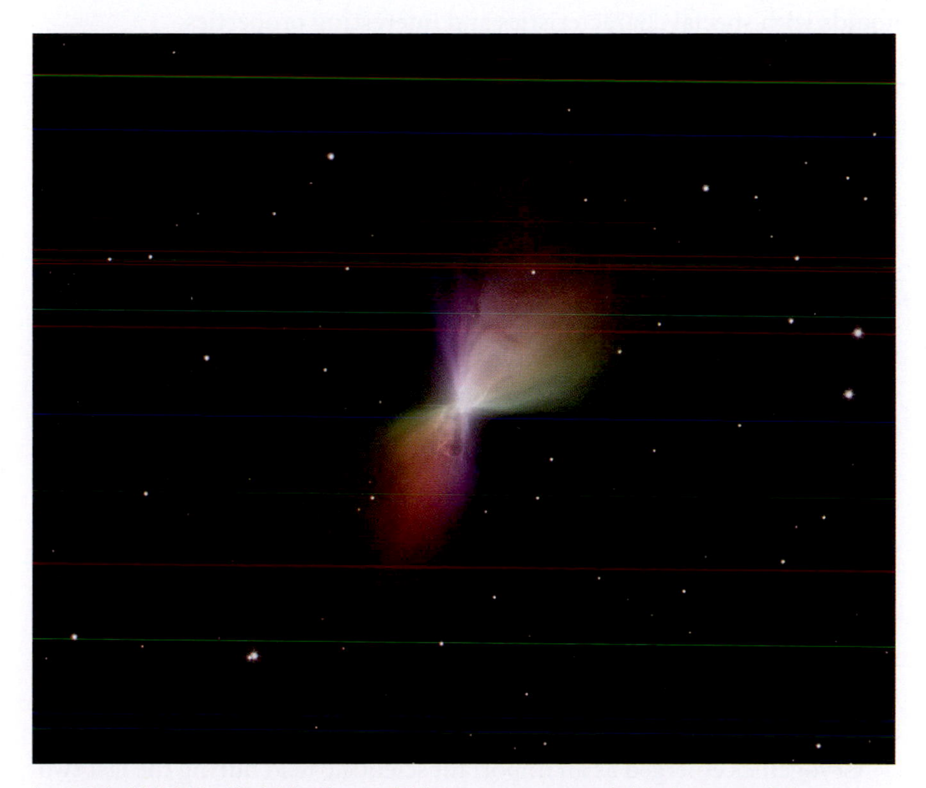

In 1998, NASA's *Hubble Space Telescope* captured this image of the Boomerang Nebula, which has a temperature of only –457.6°F (–272°C [1 K]), making it the coldest place observed so far in the universe. The high-speed, bipolar outflow of gases from the dying star are producing an expansion-cooling phenomenon. *(NASA/ESA/The Hubble Heritage Team [STScI/AURA])*

an estimated temperature of just −457.6°F (−272°C [1 K])—making it the coldest known naturally occurring place in the universe. In the 1990s, astronomers used ground-based and space-based instruments to observe the Boomerang Nebula. The unusual planetary nebula surrounds a dying star that is expelling gas during the last stages of its life. The gases flowing away from the dying star are expanding very rapidly into space. This unusual bipolar outflow at an estimated speed of 102 miles per second (164 km/s) is producing the nebula's extremely low temperature.

CRYOGENICS

While people normally think about oxygen, nitrogen, hydrogen, and helium as gases, at very low temperatures these gases become cryogenic liquids with special characteristics and interesting properties.

Cryogenics is the branch of science that deals with very low temperatures, the applications of these very low temperatures, and methods of producing them. Engineers in this field are usually concerned with practical problems, such as producing, transporting, and storing the large quantities of liquid oxygen or liquid hydrogen used for chemical rocket propellants. Engineers are also asked to design the cold-resistant equipment that allows physicists to investigate some basic properties of matter at extremely low temperatures. At extremely low, or cryogenic, temperatures, traditional research materials become very brittle. Thus, scientists must use special equipment when they explore temperature conditions at the threshold of absolute zero.

For engineers, a cryogenic temperature is typically one below −238°F (−150°C [123 K]). However, scientists often use the boiling point of liquid nitrogen at atmospheric pressure—namely −319°F (−195°C [77.4 K])—as the cryogenic temperature threshold. In extremely low-temperature research, other scientists will often treat the temperature associated with the boiling point of liquid helium—namely −452.2°F (−269°C [4.2 K])—as the threshold of the cryogenic temperature regime. Liquid helium is the most important cryogenic fluid, or cryogen, in science.

Cryogenics emerged as an important scientific field during the last two decades of the 19th century. In 1883, the Polish scientists Zygmunt Florenty Wróblewski (1845–88) and Karol Stanisław Olszewski (1846–1915) investigated the properties of liquefied air at the Jagiellonian University in Kraków and became the first researchers to produce scientifically significant quantities of liquefied oxygen and liquefied nitrogen.

Elsewhere in Europe, the Scottish scientist Sir James Dewar (1842–1923) specialized in the study of low-temperature phenomena and invented the double-walled vacuum flask that now carries his name. In 1892, Dewar began using his newly invented vacuum flask device to store liquefied gases at very low temperatures. On May 10, 1898, he used the Dewar flask and innovative regenerative cooling techniques and became the first scientist to successfully liquefy hydrogen.

The Dewar flask, or dewar, is a double-walled container with the interspace (annular space) evacuated of gas (air) to prevent the contents from gaining or losing heat by *conduction* or convection. The walls are also silvered to minimize *radiation heat transfer.* Aerospace engineers use large modern versions of this device to store a space launch vehicle's cryogenic propellants, such as liquid hydrogen and liquid oxygen (LOX).

The German engineer Carl von Linde (1842–1934) pioneered refrigeration technology toward the end of the 19th century. His efforts included developing and patenting a process for the liquefaction of atmospheric air. By 1905, Linde had perfected technologies to extract large quantities of pure oxygen and pure nitrogen from liquid air. His efforts made oxygen and nitrogen available for use in science, medicine, engineering, and industry. One consequence of Linde's work was the availability of liquid oxygen for use as a cryogenic propellant in modern liquid-fueled rocket engines.

As a result of the pioneering efforts of Dewar, Linde, and other low-temperature researchers, all temperatures down to −434.2°F (−259°C [14 K]) were attainable in laboratory environments by the beginning of the 20th century. Despite this amazing progress, one elusive goal remained. Although many scientists tried, no one had yet achieved the production of liquid helium. Then, in 1908, the Dutch physicist Heike Kamerlingh Onnes (1853–1926) developed techniques to investigate temperatures approaching absolute zero and became the first scientist to successfully liquefy helium. His research apparatus achieved a temperature of just −458.05°F (−272.25°C [0.9K])—the lowest temperature ever achieved on Earth up to that time. Onnes built upon his successful low-temperature research efforts and discovered the phenomenon of *superconductivity* in 1911. Specifically, he observed that the electrical *resistance* in a test sample of mercury (Hg) essentially vanished at −452.11°F (−268.95°C [4.2 K]). Onnes received the 1913 Nobel Prize in physics for his low-temperature research, including the production of liquid helium.

As they began to explore temperatures in the cryogenic range, scientists were challenged by how to make accurate temperature measurements.

Since mercury (Hg) freezes at –37.892°F (–38.83°C [234.32 K]), the substance became useless as the working fluid in low-temperature thermometers. Scientists and engineers turned to platinum-resistance thermometers to accurately measure temperatures down to about −423.67 °F (–253.15°C [20 K]). The platinum-resistance thermometer has a well-defined behavior of electrical resistance versus temperature. To record temperatures below −423.67°F (–253.15°C [20 K]), scientists used electrical-resistance thermometers made of certain semiconductor materials (such as doped germanium). With this approach, they were able to measure temperatures down to −457.87°F (–272.15°C [1K]) and below. Low-temperature thermometers based on such semiconductor materials require that a physical variable (such as electrical resistivity) changes in a well-known theoretical way with respect to temperature.

The production of cryogenic temperature regimes starts with the *compression* and expansion of gases. In a typical air liquefaction process, for example, engineers compress atmospheric air, causing it to heat. They then allow the compressed air to cool in a heat exchanger before rapidly expanding the gas back to atmospheric pressure. This rapid expansion (in a closed container) allows the air to cool even more—transforming a portion of the air into a liquid. Liquefied air is pale blue and contains mostly nitrogen and oxygen. At one atmosphere pressure, oxygen has a boiling point of –297.39°F (–183.0°C [90.2 K]) and nitrogen a boiling point of –321°F (–196.2°C [77 K]). The cooled gaseous portion of the expanded air returns to the mechanical compression and expansion process, while the liquid air is distilled to yield liquid oxygen, liquid nitrogen, and other atmospheric components, such as liquid argon. Scientists use gaseous helium to produce even lower cryogenic temperature environments, but several stages of mechanical compression expansion are necessary.

There are two other physical processes that scientists and engineers use to create temperatures near absolute zero. These are the Joule-Thomson (Kelvin) effect and adiabatic demagnetization. In 1852, the British scientist James Prescott Joule (1818–89) and the Scottish scientist Lord Kelvin (William Thomson) discovered that the temperature of the gas drops when a real (nonperfect) gas expands through a throttling device (such as a nozzle or a porous plug) from a high-pressure region to a low-pressure region. Using this phenomenon, engineers have developed practical refrigerators. The Joule-Thomson effect also serves as a key step in the liquefaction of gases, such as helium. When Onnes first liquefied helium in 1908, he initially chilled the helium gas by contacting it with liquid oxygen, then liquid nitrogen, and finally liquid hydrogen. Onnes then expanded the very cold helium gas through a Joule-Thomson nozzle and produced a mixture of frigid helium gas and liquid droplets.

A great deal of interesting thermodynamic theory is needed to appreciate why real gases experience Joule-Thomson effect cooling upon expansion through a throttle, while perfect (ideal) gases do not. The key point is that refrigeration devices employing the Joule-Thomson effect allow scientists to achieve temperatures approaching absolute zero.

Adiabatic demagnetization refrigerators (ADRs) are routinely used by scientists to reach temperatures below one kelvin (1 K)—that is, below–457.9°F (–272.2°C). These devices are based upon the thermodynamic process called the magnetocaloric effect. Scientists describe the magnetocaloric effect as the decrease in absolute temperature when certain substances experience adiabatic demagnetization. To utilize this interesting phenomenon, scientists first use liquid helium to cool a paramagnetic salt down to a few kelvins. Once the paramagnetic salt has been properly chilled, scientists apply a strong magnetic field. Then they thermally isolate the paramagnetic salt and reduce the applied magnetic field to zero. Because of the magnetocaloric effect, the temperature of the paramagnetic salt will drop to one kelvin (1 K), or less. Physicists regard a paramagnetic (weakly magnetic) substance as one which has an assembly of magnetic dipoles that have random orientation. Ferric ammonium sulfate ($FeNH_4(SO_4)_2 \cdot 12H_2O$) is a paramagnetic salt that NASA engineers have used to construct a cyclic magnetic cooler, or ADR.

SUPERCONDUCTIVITY

Scientists define superconductivity as the ability of a material to conduct electricity without resistance. Scientists call the temperature at which a material suddenly experiences zero (or negligible) electrical resistivity as that material's critical temperature (T_C), or the transition temperature (T_T). When cooled below its critical temperature, a material becomes a superconductor. In 1911, the Dutch physicist Heike Onnes observed that the electrical resistance in a test sample of mercury (Hg) essentially vanished at –452.11°F (–268.95°C [4.2 K]).

Other scientists soon discovered that many other metals become superconductors at temperatures near absolute zero. For example, the critical temperature of aluminum is –457.5°F (–272°C [1.18K]); tin is –453°F (–269.4°C [3.72 K]); niobium is –447°F (–266°C [7.2 K]); and lead is –443°F

This figure shows a magnet being levitated by high-temperature (copper oxide) superconducting materials that are cooled in liquid nitrogen. *(DOE)*

(–263.9°C [9.26 K]). Commercially available superconducting coils (often composed of niobium-titanium alloy [NbTi]) create the very powerful magnets that scientists use in particle accelerators and that medical engineers incorporate in magnetic resonance imaging (MRI) systems. The critical temperature of NbTi alloy is –441.7°F (–263.2°C [10 K]). Engineers plan to use superconductor materials to achieve the magnetic levitation of high-speed trains and in the development of more efficient transmission lines for electric power distribution. One difficulty is efficiently cooling metallic superconductor materials that have critical temperatures near absolute zero.

In the 1980s, the German physicist Johannes Georg Bednorz (1950–   ) and the Swiss physicist Karl Alexander Müller (1927–   ) started a systematic investigation of the electrical properties of certain ceramic materials. By 1986, they succeeded in demonstrating that lanthanum barium copper oxide (LaBaCuO, or LBCO) exhibited superconductivity at a critical temperature of –396.7°F (–238.2°C [35 K]). This was an important milestone in the development of high-temperature superconductor materials. For their discovery of superconductivity in ceramic materials, Bednorz and Müller shared the 1987 Nobel Prize in physics.

Other researchers have since identified superconductor materials that have critical temperatures near (or even above) the boiling point of liquid nitrogen (–321°F [–196.2°C {77 K}]). For example, one high-temperature copper oxide superconductor ($HgBa_2Ca_2Cu_3O_x$) has a critical temperature of –216.7°F (–138.2°C [135 K]). Ceramic materials that function at or above liquid nitrogen temperatures significantly expand the beneficial applications of superconductivity. Compared to liquid helium, liquid nitrogen is a more easily handled and much less expensive coolant.

APPLICATIONS OF CRYOGENIC LIQUIDS

Cryogenic liquids enjoy many applications. Briefly discussed in this section are numerous military, industrial, and medical uses for liquid oxygen and liquid nitrogen. Liquid hydrogen has a well-proven role in space exploration as a chemical rocket fuel and an equally significant role in future global energy infrastructure. Liquid helium plays a very special role in scientific research, space exploration, and superconductivity.

Liquid oxygen has a boiling point of –297°F (–183°C [90.2 K]) at a pressure of one atmosphere. Because oxygen is so reactive, the pale blue liquid must be stored in clean systems that are constructed of high-ignition temperature, nonreactive materials. Oxygen is often stored as a liquid, although it is used primarily as a gas. U.S. Air Force flight support personnel upload liquid oxygen (LOX) on certain military aircraft. The liquid oxygen is then converted into gaseous form to provide pure oxygen to pilots during high-altitude flight.

As an inexpensive, inert cryogen, liquid nitrogen (LN_2) enjoys many applications in research, industry, and medicine. At a pressure of one atmosphere, liquid nitrogen boils at –321°F (–196.2°C [77 K]). Like liquid oxygen, liquid nitrogen is harvested from the atmosphere by modern air-liquefaction processes. Liquid nitrogen is colorless, odorless, inert, noncorrosive, and nonflammable. This cryogen is very cold. Thus, safety procedures and personal protection equipment are essential if severe skin burns and frostbite are to be avoided.

In cryosurgery, physicians use liquid nitrogen to selectively remove unnecessary or diseased tissue. The physician employs a special precision instrument through which liquid nitrogen circulates. Upon contact with tip of the intensely cold probe (maintained at –321°F [–196.2°C {77 K}]), the target tissue is frozen and destroyed. Doctors use this generally bloodless form of surgery to treat warts or destroy certain types of tumors.

A research technician examines seeds preserved in a vat of liquid nitrogen. Scientists at the U.S. Department of Agriculture project the longevity of cryopreserved seeds to be hundreds of years; but they periodically remove them to assess their viability and vigor. *(USDA)*

Neurosurgeons use liquid nitrogen-cooled probes to treat certain brain disorders.

The practice of cryopreservation involves the use of very low temperatures to preserve indefinitely cells, tissues, blood products, skin, embryos, sperm, or similar biological materials. Scientists generally draw on liquid nitrogen to create and maintain the required low-temperature environment. Rigorous protocols are necessary for freezing and later thawing viable biological materials. These protocols vary with the biological material being preserved.

Plant physiologists use cryopreservation to store seeds of plants important to human survival. They also preserve seeds from rare and endangered plants. Similarly, veterinary scientists use cryopreservation to store the embryos and sperm of endangered animals or prized livestock. Finally, human fertility clinics help men and women who wish to use cryopreservation techniques to store their sperm or oocytes (eggs). When in vitro fertilization procedures result in more fertilized embryos than are immediately required, the physician may recommend cryopreservation to store the surplus embryos for later use.

Medical specialists use controlled-rate freezing techniques to slowly cool embryos placed in a special cryoprotectant fluid down from body temperature to –321°F (–196.2°C [77 K]). Once at this low temperature, the embryos are stored in special containers within a bath of liquid nitrogen. In 1983, the medical profession reported the first successful pregnancy from a frozen/thawed human embryo. By 2000, some 16,000 cases of assisted reproductive technology (ART) in the United States involved the use of frozen/thawed human embryos. At present, medical studies of the human offspring arising from frozen/thawed embryos have not revealed any significant increase in birth defects or physical abnormalities when compared to the pregnancy outcomes in the rest of the population. The medical profession has developed ethical guidelines concerning human embryo cryopreservation.

Liquid hydrogen has a boiling point of –423.2°F (–252.9°C [20.3 K]). Although hydrogen is the most abundant element in the universe, it is not found by itself as an element on Earth. Liquid hydrogen is often used along with liquid oxygen as the propellants in modern rocket vehicles. Liquid hydrogen also plays a role in the study and application of high-temperature superconductors. Energy engineers regard liquid hydrogen as an important energy carrier in the future.

An energy carrier is a substance or system that moves energy in a usable form from one place to another. Today, electricity is the most well known energy carrier. People use electricity to transport the energy available in coal, uranium, or falling water from generation facilities to points of application such as industrial sites and homes. An energy carrier makes the application of the energy content of the primary source more convenient. Liquid hydrogen can store energy until it is needed and can also move energy to a variety of places where it is needed.

As an extremely cold cryogen, proper equipment is needed to safely store and handle liquid hydrogen. Hydrogen gas is flammable, so care must also be exercised to avoid situations where gas leaks represent an explosive hazard. The flammable range of hydrogen in room-temperature air at one atmosphere pressure is 4 to 75 percent by volume. The flame temperature of hydrogen in air is 3,713°F (2,045°C).

The American aerospace program has successfully handled liquid hydrogen for decades. Operational and safety procedures involving the space program's use of liquid hydrogen represent an important technical legacy. These experiences can assist the growth of a global hydrogen-based energy economy this century.

Liquid helium is the most important scientific cryogen and plays a special role in research, space exploration, and superconductivity. As previously mentioned, early in the 20th century, the Dutch physicist Heike Onnes developed experimental techniques to investigate temperatures approaching absolute zero. In 1908, he became the first scientist to successfully liquefy helium. His research apparatus reached a temperature of just −458.05°F (−272.25°C [0.9K])—the lowest temperature ever achieved on Earth up to that time. Three years later, Onnes observed that at the frigid temperature of −452.11°F (−268.95°C [4.2 K]), the electrical resistance in a test sample of mercury (Hg) essentially vanished. He had discovered the important phenomenon of superconductivity

Although helium is the second-most abundant element in the universe, its existence was totally unknown prior to the late 19th century. In 1868, the British physicist Sir Joseph Norman Lockyer (1836–1920) collaborated with the French astronomer Pierre Jules César Janssen (1824–1907) and discovered the element helium through spectroscopic studies of solar prominences. Specifically, while observing a total solar eclipse, Janssen noticed a peculiar yellow line in the solar spectrum. Lockyer soon recognized that this spectral line (occurring at a wavelength of 587.49 nm) could not be produced by any element then known. Lockyer postulated the existence of a new element, which he named helium—after *helios* (ἥλιος) the sun god in ancient Greek mythology.

It was not until 1895 that the Scottish chemist Sir William Ramsay (1852–1916) finally detected helium on Earth. He made the discovery while carefully examining the uranium mineral named cleveite. The gaseous helium Ramsay detected came from the radioactive decay of uranium and was trapped within the uranium-bearing mineral. Today, helium gas is commercially recovered from natural gas deposits found primarily in Texas, Oklahoma, and Kansas. Helium makes up about 0.0005 percent of Earth's atmosphere. This trace amount is constantly being lost to space. However, the alpha decay of radioactive elements found in Earth's crust continually releases helium, some of which seeps up through cracks in the crust and replenishes the trace quantity found in the atmosphere.

As normally encountered on Earth, helium is a colorless, odorless, inert, nontoxic gas that resides at the top of the noble gas family in the periodic table (see appendix). Helium has the lowest boiling and *melting points* of all the known elements, so the substance exists as a liquid or solid only under extreme conditions of low temperature and high pressure. Unlike any other known element, helium remains a liquid down to absolute zero under normal conditions of pressure.

To appreciate the unusual properties of liquid helium, it is first necessary to recognize that helium has two natural, stable isotopes: helium-4 and helium-3. With a natural abundance (as found on Earth) of 99.999863 percent, helium-4 is by far the most prevalent form of helium. The nucleus of a helium-4 atom consists of two protons and two neutrons. The nucleus of the helium-3 atom consists of two protons and only one neutron. With a natural abundance of just 0.000137 percent, helium-3 is the extremely rare natural isotope of helium. Quantum mechanical differences in the nuclei of these two *stable isotopes* result in some very interesting behaviors as cryogenic fluids. This section primarily addresses the cryogenic behavior of helium-4; but certain properties of helium-3 as a cryogenic fluid will also be mentioned.

Helium-4 has a boiling point of −452.1°F (−268.95°C [4.2 K]) at one atmosphere pressure; helium-3 has a boiling point of −453.9°F (−269.95°C [3.2 K]) at one atmosphere pressure. Unlike any other elemental substance, helium remains a liquid down to absolute zero at normal pressures. The only way scientists are able to transform liquid helium into the solid state is to increase the pressure on the fluid while keeping it at an extremely low temperature. In 1926, the Dutch physicist Willem Hendrik Keesom (1876–1956) developed a technique to solidify helium. Scientists now understand that helium-4 transforms from a liquid to a solid at a temperature of about −457.9°F (−272.15 °C [1 K]) and a pressure of 25 atmospheres, while helium-3 experiences this transition at about 0.3 K and a pressure of 29 atmospheres.

While experimenting with liquid helium-4 at extremely low temperatures in the 1930s, scientists discovered an amazing phenomenon, called superfluidity. They observed that between 4.22 K and 2.177 K and one atmosphere pressure, the isotope helium-4 was a very cold, colorless liquid. (Recall that 4.22 K is the boiling point of helium-4 at one atmosphere pressure.) Scientists called this form of liquid helium-4, helium I (often abbreviated as He-4I). Like other cryogenic liquids, helium I boils when heated and contracts when the temperature is lowered. This behavior continues until the helium I reaches 2.177 K.

At 2.177K, helium I experiences a remarkable physical discontinuity. The *thermal conductivity* of the cryogenic liquid increases dramatically and its *viscosity* becomes zero. Scientists refer to the temperature associated with this transition as the lambda point; they call liquid helium

SUPERFLUIDS

Scientists define *superfluid* as referring to a zero viscosity liquid that flows with no inner friction. Physicists further note that a liquid becomes a superfluid when all its atoms or molecules have been cooled down (or condensed) to the point where they all occupy the same quantum state. Toward the end of the 1930s, the Russian physicist Pyotr Kapista (1894–1984) began a series of experiments in low-temperature physics involving the study of liquid helium. His work led to the discovery (in 1937) of superfluidity involving the form of liquid helium-4, called helium II. He received a portion of the 1978 Nobel Prize in physics for his pioneering low-temperature research. Kapista's work also inspired another Russian physicist, Lev Davidovich Landau (1908–68). Landau performed important theoretical studies concerning the unusual behavior of liquid helium at near absolute zero. He received the 1962 Nobel Prize in physics for his pioneering theories of condensed matter, especially as related to the superfluid behavior of liquid helium-4.

Such research efforts clearly indicated that classical physics was inadequate to explain the phenomenon of superfluidity. Superfluids exhibit a variety of unusual, nonclassical behaviors—including flow through very small cracks and creep flow up and out of containers. Very cold liquids that exhibit superfluidity are often termed quantum fluids because physicists must use advanced forms of quantum mechanics to explain their behavior. Today, scientists typically refer to a superfluid as a quantum fluid with perfect microscopic (that is atomic-level) order or zero entropy.

In 1996, three American physicists, David M. Lee (1931–), Douglas D. Osheroff (1945–), and Robert C. Richardson (1937–), shared the Nobel Prize in physics for their discovery that liquid helium-3 exhibited superfluidity at a temperature of about 0.0025 K. Research suggests that as a quantum fluid liquid helium-3 possesses a significantly more complicated nuclear structure than the quantum fluid, liquid helium-4. For example, liquid helium-3 is anisotropic—meaning the quantum fluid displays different physical properties in different spatial directions. The British-American physicist Sir Anthony James Leggett (1938–) received the 2003 Nobel Prize in physics for his theoretical work concerning the superfluidity of helium-3.

cooled below the lambda point a superfluid. Also termed helium II (He-4II), this form of liquid helium is unlike any other known substance. Below the lambda point, helium II does not boil and exhibits superfluid behavior. This means helium II can flow through microscopic channels and very tiny capillaries and creep out of supposedly leak-tight containers. Cryogenic engineers discovered that they had to design very special containers to store helium II; otherwise, the superfluid would creep along the container's surfaces and sneak through valves and microscopic cracks until it reached a warmer region, where it would evaporate.

Liquid helium plays an essential role in modern science, including the operation of large-particle accelerators that rely on superconductors and the cooling of special instruments on spacecraft. In 1989, NASA successfully launched the *Cosmic Background Explorer (COBE)* into orbit around Earth. Instruments on this scientific spacecraft provided the scientific community with its first definitive measurement of the cosmic microwave background—thereby confirming the big bang theory of the origin of the universe. The *COBE* spacecraft carried a cryostat filled with 174 gallons (660 L) of liquid helium. The liquid helium provided a stable −457.15°F (−271.75°C [1.4 K])—environment for two of the spacecraft's instruments: the far infrared absolute spectrometer (FIRAS) and the diffuse infrared background experiment (DIRBE). On September 21, 1990, after 306 days of successful cryogenic operations, the last of the superfluid helium contained in the dewar was consumed, and the initial phase of this important space mission came to an end. The American scientists John C. Mather (1946–   ) and George F. Smoot (1945–   ) received the 2006 Nobel Prize in physics for their discovery of the blackbody form and anisotropy of the cosmic microwave background.

BOSE-EINSTEIN CONDENSATE

The Bose-Einstein condensate (BEC) is the state of matter in which extremely cold atoms attain the same quantum state and behave like one large superatom. The fascinating phenomenon is named after the Indian mathematical physicist Satyendra Nath Bose (1894–1974), who collaborated with Albert Einstein and made important mathematical contributions to quantum mechanics. Expanding on Bose's work regarding particles of light, Einstein postulated (in 1925) that if a gas was cooled to an extremely cold state (much less than one-millionth of a kelvin), then the great majority of the atoms in the frigid sample would settle into the lowest possible energy state in the containment vessel. The process resembles the formation of liquid drops when a vapor is cooled; thus, scientists call the anticipated superatom product the Bose-Einstein condensate.

Guided by the uncertainty principle of the German theoretical physicist Werner Karl Heisenberg (1901–76), physicists in the late 1920s began using wave equations to develop a quantum mechanical description of matter on the atomic scale. In the case of the Bose-Einstein condensate, the individual wave equations (describing the velocity and position of an atom) merge and become indistinguishable from one another. In the process of Bose-Einstein condensation, the wave function of each atom gets in phase with the wave functions of the other very cold atoms in the sample. As experiments at the end of the 20th century revealed, scientists could actually observe on a macroscopic scale the quantum mechanical behavior of the atoms that make up the condensate. Like superconductivity and superfluidity, the BEC provided scientists a rare macroscopic window into the strange, often bizarre microscopic world of quantum mechanics.

It took approximately seven decades before scientists could experimentally follow up on Einstein's predictions about the formation of a giant superatom under extremely low temperature conditions. In 1925, no scientist seriously believed that temperatures approaching to within one-billionth of a degree of absolute zero would ever be achieved in the laboratory. Because of such apparently prohibitive experimental barriers, the notion of the BEC was basically placed in the scientific attic and remained there for decades.

Starting in the 1970s, progress in advanced refrigeration and laser technologies encouraged scientists to revisit Einstein's original notion of the BEC. Over the next two decades, scientists learned how to use lasers to both cool and trap atoms. After much hard work and innovation, the laboratory stage was set for a major breakthrough in materials science.

In June 1995, a team of scientists working at JILA succeeded in creating a tiny, but observable, amount of a new, extreme state of matter called the Bose-Einstein condensate. JILA (formerly called the Joint Institute for Laboratory Astrophysics) is a joint institute of the University of Colorado at Boulder and the National Institute of Standards and Technology (NIST). The American physicists Carl E. Wieman (1951–) and Eric A. Cornell (1961–) successfully cooled a minuscule sample of about 2,000 rubidium atoms to 0.00000002 degree above absolute zero (that is, 20 nK). As predicted by Einstein 70 years earlier, the extreme, low-temperature conditions caused the individual rubidium atoms in this very cold gas to condense into a superatom. The condensate then behaved (for about 10 seconds) as a single entity. The accompanying figure shows three-dimensional successive snapshots in which the rubidium atoms condensed from less dense red, yellow, and green areas into very dense blue to white areas. The American scientists achieved the rare opportunity of observing on a macroscopic level how quantum mechanical waves extended across the frigid sample of rubidium atoms and formed the Bose-Einstein condensate.

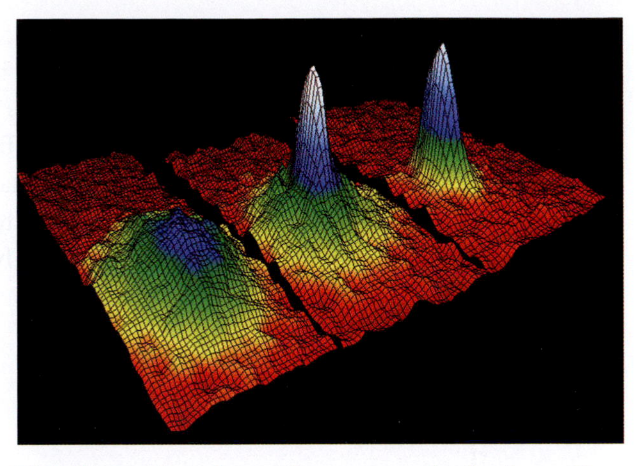

This illustration shows three stages in the formation of a Bose-Einstein condensate (BEC). In 1995, the American Nobel laureates Eric Cornell and Carl Wieman cooled a gas containing about 2,000 rubidium atoms to an incredibly low temperature of 0.00000002 Kelvin (20 nK). The three snapshotlike images depict progressively colder conditions (warmest on left, coldest on right). In the image on the right (tallest peak and very dense blue to white areas), the BEC is fully formed and behaving as a single superatom. *(NIST/JILA/CU-Boulder)*

Several months after Wieman and Cornell produced the world's first BEC in a dilute gas of rubidium, the German physicist Wolfgang Ketterle (1957–)—working at the Massachusetts Institute of Technology (MIT) in Cambridge, Massachusetts—produced a BEC with sodium atoms. Subsequent activities by Ketterle's team at MIT resulted in BEC's containing about 10 million atoms. Ketterle also produced a stream of tiny BEC drops that descended under the influence of gravity.

The international scientific community quickly recognized the significance of these experiments. The 2001 Nobel Prize in physics was awarded to Cornell, Ketterle, and Wieman for "the achievement of Bose-Einstein condensation in dilute gases of alkali atoms and for early fundamental studies of the properties of the condensates."

Most scientists think it is too early to speculate about possible practical applications of BECs. For now, most low-temperature physicists remain focused on further exploring the properties and characteristics of this extreme state of matter. However, a few of them have suggested that BECs will eventually trigger a revolution in such areas as nanotechnology and ultra-precise instrumentation. One interesting analogy provides a hint as to the future importance of the BEC. During the Nobel Prize presentations in 2001, it was suggested that the BEC resembles a "special laser beam" in which the collection of atoms (matter) all behaving in unison have replaced the collection of highly organized photons characteristic of a traditional optical laser.

CHAPTER 8

Antimatter

This chapter introduces the scientific aspects and applications of anti- matter. Although commonly associated with science fiction, antimatter is a real, physical phenomenon that serves an important role in modern physics research, astronomy, and medicine.

THE DISCOVERY OF ANTIMATTER

Scientists define antimatter as matter in which the ordinary nuclear particles (such as electrons, protons, and neutrons) have been replaced by their corresponding antiparticles (that is, positrons, antiprotons, and antineutrons.) The nuclear physics shorthand used to identify an antiparticle is to place a bar (line) over the top of the letter representing the ordinary matter nuclear particle. For example, the antiparticle of the electron (e) could be identified by the e-bar symbol ($\bar{e}$), although this antiparticle is more commonly called the positron and noted by the symbol e^+. Similarly, scientists usually specify the antiparticle of the neutrino (ν) by using the nu-bar ($\bar{\nu}$) symbol and the antiparticle of the proton (p) by using the p-bar ($\bar{p}$) symbol.

The possibility of antimatter was overlooked by scientists until the emergence of nuclear science and quantum mechanics in 20th century. In 1928, the British theoretical physicist Paul Adrien Maurice Dirac

(1902–84) developed a relativistic wave equation that correctly described the electron's spin (intrinsic angular momentum). However, Dirac's equation also had a negative energy solution, which he soon interpreted (in 1930) as the possible existence of antimatter—specifically a positively charged electron. Dirac coshared the 1933 Nobel Prize in physics with the Austrian physicist Erwin Schrödinger (1887–1961) for the discovery of new productive forms of quantum theory.

While using a cloud chamber in 1932 to study cosmic-ray particles, the American physicist Carl David Anderson (1905–91) discovered the positron. This discovery transformed the overall understanding of matter and essentially doubled the contents of the material universe. Anderson continued to investigate subatomic particles and in 1933 was able to demonstrate that positrons were emitted from gamma rays in a process now called pair-production. Anderson coshared the 1938 Nobel Prize in physics. His Nobel lecture was entitled "The production and properties of positrons."

In the process of pair-production, a high-energy gamma ray interacts with matter and disappears and an electron (e^-)–positron (e^+) pair appears in its place. Pair-production generally takes place in the vicinity of the strong electrostatic (coulomb) field of an atomic nucleus. It is a threshold process requiring that the incident gamma ray have an energy level of at least 1.02 MeV, which is the rest mass equivalent energy of the electron-positron particle pair. The probability of pair-production increases as the photon energy exceeds 1.02 MeV and becomes the dominant mechanism for gamma-ray interaction with matter above about 5 MeV. Every positron created in pair-production loses its kinetic energy by causing ionization in the absorbing medium. Spent of its kinetic energy, the positron then interacts an ordinary electron, converting the electron and itself into a pair of 0.51 MeV annihilation radiation gamma rays. Today, the most commonly encountered antiparticle is the positron. As discussed later in the chapter, positrons play a major role in modern nuclear medicine.

In 1955, the Italian-American physicist Emilio Gino Segrè (1905–89) and the American physicist Owen Chamberlain (1920–2006) collaborated at the University of California, Berkeley, and their mutual research activities resulted in the discovery of the antiproton. They coshared the 1959 Nobel Prize in physics for their important effort. Today, scientists use the standard model of matter to state that the antiproton consists of two up antiquarks and one down antiquark.

POWERFUL ACCELERATOR DISCOVERS EXOTIC ANTIMATTER

In March 2010, an international team of scientists described how they used the Relativistic Heavy Ion Collider (RHIC) at the Brookhaven National Laboratory in New York to discover the heaviest known antinucleus, which also is the first antinucleus found that contains an antistrange quark. The scientists carefully investigated collisions of extremely energetic gold ions. The new antinucleus is a negatively charged state of matter containing an antiproton, an antineutron, and an antilambda particle. The exotic sample of antimatter is also the first observed to contain an antistrange quark. The researchers suggest that this particular discovery paves the way to a new frontier in physics. Scientists call the exotic chunk of antimatter the antihypertriton. This name infers a nucleus of antihydrogen containing one antiproton, one antineutron, and one heavy relative of the antineutron, an antilambda hyperon.

Scientists suggest that all terrestrial nuclei consist of protons and neutrons, which in turn contain only up (u) and down (d) quarks (see chapter 3). Chemists have arranged the periodic table of elements according to the number of protons, which determines the chemical properties of each element (see appendix). However, physicists have developed a more complex, three-dimensional (3-D) chart of the nuclides. On this 3-D chart, the number of protons, also called the atomic number (Z), is plotted against the number of neutrons (N) in a particular nuclide. There is also a third axis, which represents a quantum number called the strangeness (S). The strangeness depends on the presence of strange quarks. Scientists call nuclei containing one or more strange quarks hypernuclei.

The strangeness has a value of zero for all naturally occurring matter found on Earth. However, scientists suspect that the strangeness might be nonzero for matter found in the core of collapsed stars. For ordinary matter (that is, matter not containing strange quarks), the strangeness value is zero, and the plot collapses into a two-dimensional depiction of atomic number (Z) versus neutron number (N). Previously observed hypernuclei appeared above the Z-N plane. The recent discovery of antihypertriton marks the first entry below the Z-N plane. The more scientists learn about the behavior of antimatter in this century, the better they will be able to explain what happened in the very early universe when almost equal amounts of matter and antimatter formed. The intensely energetic nucleus-nucleus collisions at RHIC provide scientists a fleeting glimpse at the conditions after the big bang, when the universe was just a few microseconds old.

One year after the antiproton was discovered, other scientists at the University of California, Berkeley, used a particle accelerator (called the Bevatron) to pass a beam of antiprotons through ordinary matter and created antineutrons, as some of the antiprotons in the beam exchanged their negative charges with nearby positively charged protons. Scientists were able to detect the antineutrons. As scientists continued to explore the inner secrets of matter with more powerful particle accelerators, the collection of antiparticles in the overall nuclear particle zoo also increased dramatically.

ANTIPROTONS AND ANTIHYDROGEN

Physicists using high-energy accelerators often make millions and millions of antiprotons, which they keep trapped in magnetic fields. Recently, scientists have learned how to combine positrons and antiprotons carefully to form antimatter hydrogen atoms, sometimes referred to as antihydrogen. An antimatter hydrogen atom (or antihydrogen) consists of a negatively charged antiproton as the nucleus surrounded by a positively charged orbiting positron. Scientists have to use special magnetic traps to hold antihydrogen, or else it would destroy itself in a flash of radiant energy upon contact with ordinary matter.

The sprawling accelerator complex (highlighted in red) at the Fermi National Laboratory (Fermilab) in Batavia, Illinois accelerates protons and antiprotons close to the speed of light. The Tevatron collider, some four miles (6.4 km) in circumference, produces millions of proton-antiproton collisions per second. Sophisticated detectors record the collisions, allowing scientists to hunt for telltale signs of new nuclear particles and subatomic phenomena. *(DOE/Fermilab)*

CERN

CERN, the European Organization for Nuclear Research, is one of the world's major nuclear research centers. Founded in 1954, the CERN Laboratory sits astride the Franco-Swiss border near Geneva, Switzerland. International teams of scientists from member states use powerful particle accelerators and sophisticated detector complexes to explore the basic constituents of matter. The centerpiece machine is the Large Hadron Collider (LHC)—the world's largest and most powerful particle accelerator. The LHC consist of a 16.8-mile (27-km) ring of superconducting magnets with a number of accelerating structures to boost the energy of particles along the way. The magnets are kept at a frigid temperature of −456°F (−271°C) by liquid helium. When operating at full power, trillions of protons will race around the LHC accelerator ring at 99.9999991 percent of the speed of light.

Normal matter and antimatter mutually annihilate each other upon contact and are converted into pure energy, called annihilation radiation. When an electron and a positron annihilate each other, their rest masses are converted into pure energy and are replaced by two characteristic 0.51 MeV gamma rays that depart from the scene of the annihilation event in exactly opposite directions. These characteristic 0.51 MeV gamma rays are very useful in determining the presence (however fleeting) of positrons in the environment. Although extremely small quantities of antimatter (primarily antihydrogen) have been produced in laboratories, significant quantities of antimatter have yet to be produced or observed elsewhere in the universe.

On November 17, 2010, officials at CERN announced that a team of collaborating scientists had successfully produced and magnetically trapped antihydrogen atoms. Although antihydrogen atoms had been produced at CERN as early as 1995 and at Fermilab in 1997, the energetic nature of those antihydrogen atoms prevented the researchers from cooling, trapping, and successfully confining the precious bits of human-made antimatter for subsequent analysis. Scientists around the world regard the announcement from CERN as a major step in the development of advanced experimental techniques that will allow them to further explore the basic difference between matter and antimatter.

THE UNIVERSE'S MISSING ANTIMATTER

Since scientists have observed matter and antimatter being created in equal amounts in various highly energetic particle experiments, they postulated that after the big bang, when the universe was very hot and extremely dense, equal amounts of matter and antimatter were formed from the available energy. However, scientists have deduced that all matter observed in the universe today is only about one-billionth of the amount of matter that existed during the very early universe. They base this startling conclusion on numerous astronomical searches for telltale annihilation radiation signatures by various orbiting spacecraft (including NASA's *Fermi Gamma-ray Space Telescope* launched in June 2008) and the results of countless nuclear physics experiments performed in laboratories on Earth. Apparently, as the universe expanded and cooled, almost every primordial matter particle collided with an ancient antimatter particle, and the two turned into energetic gamma rays, known as annihilation radiation. But roughly one out of every billion or so matter particles

ANTIMATTER COSMOLOGY

Antimatter cosmology is a cosmological model proposed by the Swedish scientists Hannes Alfvén and Oskar Benjamin Klein (1894–1977) in the 1960s as an alternative to the big bang model. In their model, the scientists assumed that early universe consisted of a huge, spherical cloud, called a metagalaxy, containing equal amounts of matter and antimatter. As this cloud collapsed under the influence of gravity, its density increased and a condition was reached in which matter and antimatter collided—producing large quantities of annihilation radiation. The radiation pressure from the annihilation process caused the universe to stop collapsing and to expand. In time, clouds of matter and antimatter formed into equivalent numbers of galaxies and antigalaxies. An antigalaxy is a galaxy composed of antimatter.

Space-age astronomical observations of the heavens indicated many technical difficulties with the Alfvén-Klein cosmological model. For example, no observational evidence has yet been found to indicate the presence of large quantities of antimatter in the universe. If the postulated antigalaxies actually existed, large quantities of annihilation radiation (gamma rays) would certainly be emitted at the interface points between the matter and antimatter regions of the universe.

survived. The surviving matter particles now make up the galaxies, stars, planets, and all things on Earth.

Scientists now treat this amazing survival story and the absence of large quantities of antimatter in the observable universe as being indicative of subtle differences between matter and antimatter—differences that extend beyond the obvious difference involving opposite charges. There is a small difference between the way matter and antimatter interact.

This subtle, but very important, difference was first observed during particle experiments performed at the Brookhaven National Laboratory in 1964 by a research team led by the American physicists James Watson Cronin (1931–) and Val Logsdon Fitch (1923–). They observed what physicists called CP violation and received the 1980 Nobel Prize in physics for this great discovery.

Particle physicists had previously assumed that a basic symmetry existed in nature. The concept of CP symmetry represents a combination of charge conjugation symmetry (C symmetry) and parity symmetry (P symmetry). What this somewhat daunting nomenclature means is that by assuming CP symmetry, the nuclear physicists are postulating that the basic laws of physics would remain identical if normal particles were interchanged with their antiparticles.

By using more powerful accelerators and closely examining the behavior of antiatoms, such as antihydrogen, physicists are now trying to examine additional sources of CP violation. Future experiments at CERN, Fermilab, and other major research facilities should provide scientists a better insight into why all the bulk quantities of antimatter in the universe disappeared, while a tiny fraction of the original ordinary matter survived and became today's observable universe.

POSITRON EMISSION TOMOGRAPHY

One important use of antimatter is in nuclear medicine to help physicians obtain useful images of various parts of a living patient's body, such as the brain or heart muscle. Nuclear medicine technicians use short-lived positron (β^+) emitters, such as carbon-11, nitrogen-13, oxygen-15, or fluorine-18, in a process known as positron emission tomography (PET). The term *tomography* comes from the Greek words *tomos* (τομος), which means "to cut or section," and *grapho* (γραφω), which means "to write or record." The radioisotope carbon-11 has a half-life of 20.4 minutes and decays by positron emission into stable boron-11. The radioisotope nitrogen-13 has a

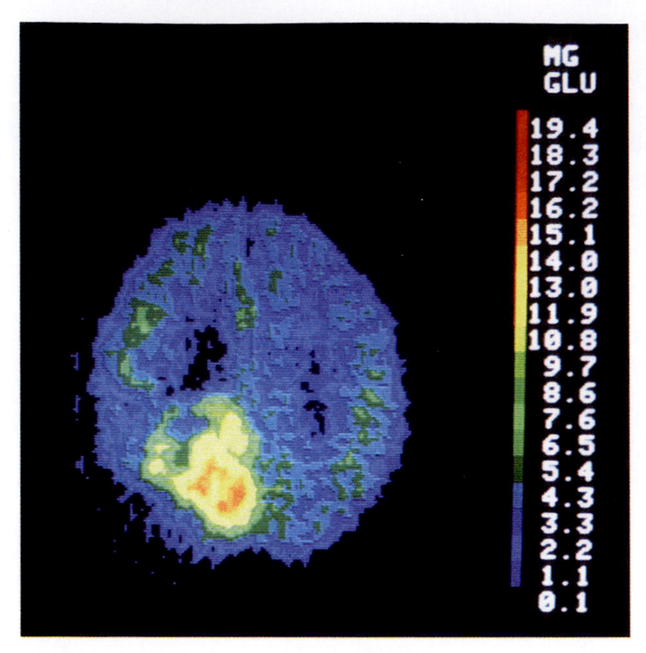

This image shows a positron emission tomography (PET) scan of a 62-year-old man with a brain tumor. Physicians use such PET images to differentiate between healthy and cancerous tissue. A tumor uses more glucose than normal tissue and appears brighter in an image. The irregular bright yellow and orange area in the lower left portion of the image indicates the location of the tumor, which metabolizes glucose faster than normal cells. The color scale on the right corresponds to milligrams of glucose (mg glu). *(National Cancer Institute)*

half-life of 9.97 minutes and decays by positron emission into the stable isotope carbon-13. Similarly, the positron-emitting radioisotope oxygen-15 has a half-life of 122 seconds and fluorine-18 a half-life of 1.82 hours.

In PET, the nuclear medicine technician places a positron-emitting radionuclide (such as the ones identified previously) into a suitable chemical compound, which once introduced into the body selectively migrates to specific organs. Medical diagnosis depends on the important physical phenomenon of pair-production, during which a positron and an ordinary electron annihilate each other and two gamma rays of identical energy (typically 0.511 MeV each) depart the annihilation radiation spot in exactly opposite directions. Radiation sensors mounted on a ring around the patient detect the pair of simultaneously emitted gamma rays and reveal a line on which the annihilation reaction occurred within the organ under study. When a large number of coincident gamma ray pairs have been detected, physicians use a computer system to reconstruct the PET image of an organ as it metabolizes the positron-emitting radioactive compound. For example, by attaching a positron emitter to a protein or glucose molecule and allowing the body to metabolize it, doctors can study the functional aspects of a living organ, such as the brain.

ANTIMATTER FOR INTERSTELLAR PROPULSION

An interesting variety of matter-antimatter propulsion schemes (sometimes referred to as photon rockets) have been suggested for interstellar travel.

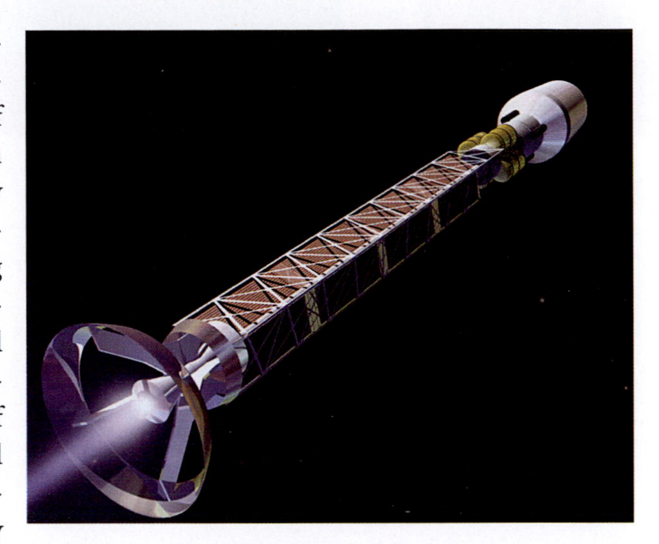

This is an artist's concept of a hypothetical antimatter space propulsion system. *(NASA/MSFC)*

Should engineers be able to manufacture and contain significant quantities (e.g., milligram to kilogram) of antimatter, such as antihydrogen, in the later decades of the 21st century photon rocket concepts might represent an interesting way of propelling robot probes to neighboring star systems. One challenging technical problem for future aerospace engineers would be the safe storage of antimatter propellant in a normal matter spacecraft. Another engineering challenge would be properly shielding the interstellar probe's payload from exposure to the penetrating, harmful annihilation radiation (intense gamma rays) released when normal matter and antimatter collide.

The basic principle of operation of the photon rocket is that it uses matter and antimatter as propellant. When combined in equal amounts, the fuel particles annihilate each other, releasing an equivalent amount of energy in the form of hard nuclear radiation (typically energetic gamma rays). In one design approach, the gamma rays are collected and emitted in a collimated beam out the back of the vessel, providing a forward thrust.

A few aerospace engineers have suggested that the (hypothetical) photon rocket represents the best theoretical propulsion system that humans' current understanding of physics will permit. Anticipated cruising speeds in interstellar space range from 0.1 c to 0.99 c, where c is the speed of light. Potential applications of the photon rocket vary from robot interstellar probes (including self-replicating systems) to human-crewed, large space arks.

Despite the ease with which antimatter powers starships in science fiction, many major technological barriers must be overcome before this type of propulsion system ever becomes a reality. Unfortunately, the projected availability of this type of advanced propulsion system lies well beyond the distant technical horizon and remains perhaps centuries away, if ever.

CHAPTER 9

Beyond Einstein

"Knowledge is limited. Imagination encircles the world."
—Albert Einstein

This chapter stretches the reader's imagination by extending the consequences of Einstein's work beyond the present and anticipating future scientific developments. Following his introduction of relativity (special and general) in the early 20th century, Einstein, like many other scientists after him, began searching for a theory of everything (TOE)—a theory that would unite the large-scale interpretation of the universe characterized by general relativity with the successful description of matter at the atomic and subatomic scales provided by quantum mechanics. Interesting concepts such as gravitational lensing, gravitational waves, and even wormholes, have emerged from the new interpretation of gravity provided by Einstein's general relativity. Today, physicists often use the general term *quantum gravity* to describe the new theories they are suggesting to unify gravitation with the other three fundamental forces of physics, represented within the standard model of quantum mechanics. As apparent throughout the chapter, this part of the scientific frontier is very rich with bold new ideas and speculative models, such as *string theory.*

EINSTEIN AND RELATIVITY THEORY

Relativity theory is the theory of space and time developed by Albert Einstein, which served as one of the pillars of physics in the 20th century. The other pillar is Max Planck's quantum theory. Physicists often discuss Einstein's theory of relativity in two general categories: the special theory of relativity, which he first proposed in 1905, and the general theory of relativity, which he presented in 1915.

The special theory of relativity is concerned with the laws of physics as seen by observers moving relative to one another at constant velocity—that is, by observers in nonaccelerating or inertial reference frames. Special relativity has been well demonstrated and verified by many types of experiments and observations.

Einstein proposed two fundamental postulates in formulating special relativity: (1) First Postulate of Special Relativity: The speed of light (c) has the same value for all (inertial-reference-frame) observers, regardless and independent of the motion of the light source or the observers.

(2) Second Postulate of Special Relativity: All physical laws are the same for all observers moving at constant velocity with respect to one another.

The first postulate appears contrary to a person's everyday Newtonian mechanics experience. Yet, the principles of special relativity have been more than adequately validated in experiments. Using special relativity, scientists can now predict the space-time behavior of objects traveling at speeds from essentially zero up to those approaching that of light itself. At lower velocities, the predictions of special relativity become identical with classical Newtonian mechanics. However, when scientists deal with objects moving close to the speed of light, they must use relativistic mechanics.

Nuclear physicists know that no particle (with mass) can move with a speed faster than the speed of light in a vacuum. They also know that there is no limit on the energy a particle can attain. In powerful modern particle accelerators, nuclear particles typically whirl around at speeds very close to the speed of light. Because of special relativity, when a particle is traveling at say 99.9998 percent of the speed of light, any significant increase in energy results in only a minimal increase in velocity. In theory, an infinite amount of energy would be needed to make the particle travel at precisely the speed of light.

Special relativity introduces some very interesting physical consequences. The first interesting relativistic effect is called time dilation. Simply stated—with respect to a stationary observer/clock—time moves more slowly on a moving clock/system. This unusual relationship is described by the equation $\Delta t = (1/\beta)\ \Delta T_p$, where Δt is called the time dilation (the apparent slowing down of time on a moving clock relative to a stationary clock/observer) and ΔT_p is the "proper time" interval as measured by an observer/clock on the moving system. The term β is defined as: $\beta = \sqrt{[1 - (v^2/c^2)]}$, where v is the velocity of the object and c is the velocity of light.

It is interesting to explore the time-dilation effect with respect to a hypothetical starship flight from the solar system. The scenario starts with twin brothers, Astro and Cosmo, who are both astronauts and are currently 25 years of age. Astro is selected for a special 40-year-duration starship mission, while Cosmo is selected for the ground control team. This particular starship, the latest in the fleet, is capable of cruising at 99 percent of the speed of light (0.99 c) and can quickly reach this cruising speed. During this mission, Cosmo, the twin who stayed behind on Earth, has aged 40 years. For the purpose of this thought experiment, Earth is taken as the fixed or stationary reference frame relative to the starship. However, Astro, who has been on board the starship traveling the local portions of the Milky Way galaxy at 99 percent of the speed of light for the last 40 Earth-years, has aged just 5.64 Earth-years! When he returns to Earth from the starship mission, he is a little more than 30 years old, while his twin brother, Cosmo, is now 65 and has retired in Florida. Obviously, starship travel (if future engineers can overcome some extremely challenging technical barriers) also presents some very interesting social problems.

The time-dilation effects associated with near-light speed travel are real, and they have been observed and measured in a variety of modern experiments. All physical processes (chemical reactions, biological processes, nuclear-decay phenomena, and so on) appear to slow down when in motion relative to a fixed or stationary observer/clock.

Another interesting effect of relativistic travel is length contraction. Physicists first define an object's proper length (L_p) as its length measured in a reference frame in which the object is at rest. Then, the length of the object when it is moving (L)—as measured by a stationary observer—is always smaller, or contracted. The relativistic length contraction is given by: $L = \beta\ (L_p)$. This apparent shortening, or contraction, of a rapidly moving object is seen by an external observer (in a different inertial reference frame) only in the object's direction of motion. In the case of a starship traveling at near-light speeds, to observers on Earth, such a hypothetical vessel would appear to shorten, or contract, in the direction of flight. If an alien starship was 0.62 mile (1 km) long (at rest) and entered humans'solar system at an encounter velocity of 90 percent of the speed of light (0.9 c), then a terrestrial observer would see an alien starship that appeared to be about 1,427 feet (435 m) long. The (hypothetical) aliens on board and all their instruments (including tape measures) would look contracted to external observers but would not appear any shorter to those on board the ship (that is, to observers within the moving reference frame). If this alien starship was really burning rubber at a velocity of 99 percent of the speed of light (0.99 c), then its apparent contracted length to an observer on Earth would be about 462 feet (141m). If, however, this vessel was just a slow interstellar freighter that was lumbering along at only 10 percent of the speed of light (0.1 c), then it would appear about 3,265 feet (995 m) long to an observer on Earth.

Special relativity also influences the field of dynamics. Although the rest mass (m_o) of a body is invariant (does not change), its relative mass increases as the speed of the object increases with respect to an observer in another fixed or inertial reference frame. An object's relative mass is given by: $m = (1/\beta)\ m_o$. This simple equation has far-reaching consequences. As an object approaches the speed of light, its mass becomes infinite. Since things cannot have infinite masses, physicists conclude that material objects cannot reach the speed of light in a vacuum. This is called the speed-of-light barrier, which appears to limit the speed at which interstellar travel can occur. From the theory of special relativity, scientists now conclude that only a zero-rest-mass particle, such as a photon, can travel at the speed of light. There is another major consequence of special relativity that has greatly affected daily lives—the equivalence of mass and energy from Einstein's very famous formula, $E = \Delta m\ c^2$, where E is the energy equivalent of an amount of matter (Δm) that is annihilated or converted completely into pure energy and c is the speed of light. This simple yet powerful equation explains where all the energy in nuclear fission or nuclear fusion comes from.

In 1915, Einstein introduced his general theory of relativity. He used this development to describe the space-time relationships developed in special relativity for cases where there was a strong gravitational influence such as white dwarf stars, neutron stars, and black holes. One of Einstein's conclusions was that gravitation is not really a force between two masses (as postulated in Newtonian mechanics), but rather arises as a consequence of the curvature of space-time. In Einstein's perspective of a four-dimensional universe (x, y, z and time), space-time becomes curved in the presence of matter—especially very massive, compact objects.

The fundamental postulate of general relativity is also called Einstein's principle of equivalence: The physical behavior inside a system in free-fall is indistinguishable from the physical behavior inside a system far removed from any gravitating matter (that is, the complete absence of a gravitational field).

Numerous experiments have been performed to confirm the general theory of relativity. These experiments have included observation of the bending of electromagnetic radiation (starlight and radio wave transmissions from various spacecraft missions) by the Sun's immense gravitational field and recognizing the subtle perturbations (disturbances) in the orbit (at perihelion—the point of closest approach to the Sun) of the planet Mercury as caused by the curvature of space-time in the vicinity of the Sun. Einstein's general relativity theory also predicts that the rotation of a massive celestial object (such as a black hole or a neutron star) would drag the local space-time around with it, ever so slightly. Astrophysicists call this subtle, difficult-to-measure phenomenon frame-dragging.

Einstein's general relativity theory transformed space from the classic Newtonian concept of vast emptiness with nothing but the invisible force of gravity to rule the motion of matter to an ephemeral fabric of space-time, which enwraps and grips matter and directs its course through the universe. Einstein postulated that the fabric of space-time spans the entire universe and is intimately connected to all matter and energy within it.

At this point, a person may wonder how this dramatic change in thinking explained the motion of the planets or the orbits of the Moon and artificial satellites around Earth. Theoretically, when a mass sits in the fabric of space-time, it will deform it, changing the geometry (shape) of space and altering the passage of time around it. It may be helpful to imagine the space-time fabric as a sturdy, but flexible, sheet of rubber stretched out to form a plane upon which an object like a baseball is placed. The ball causes the sheet to bend. In Einstein's theory of gravity, a mass causes the fabric of space-time to bend in a similar manner. The more massive the object, the deeper is the dip in the fabric of space-time. Physicists sometimes refer to this mass-induced dip in the space-time continuum as a gravity well. A black hole has an infinitely deep gravity well.

In the case of the Sun, Einstein reasoned that because of its mass the space-time fabric would curve around it, creating a dip in space-time. When the planets (as well as asteroids and comets) travel across the space-time fabric, they respond to this Sun-caused dip and travel around the massive object by following the curvature in space-time. As long as they never slow down, the planets would maintain regular orbits around the Sun—neither spiraling in toward it nor flying off into interstellar space.

GRAVITATIONAL LENSING

Einstein's general theory of relativity implies that a massive celestial object can deform the local space-time and bend the path of light coming from more distant sources. Astronomers have found this phenomenon very useful in surveying distant galaxies and in mapping the distribution of dark matter. The mass-induced distortion of space-time acts as a lens, focusing the light from distance objects toward the observer and causing this light to appear distorted and magnified, or even as multiple images. Observations suggest that accumulations of both ordinary matter and dark matter influence gravitational lensing.

The illustration on page 172 shows how gravitational lensings by foreground galaxies can influence the appearance of much more distant background galaxies. Studies in 2010 suggest that as many as 20 percent of the most distant galaxies currently detected by advanced astronomical systems such as NASA's *Hubble Space Telescope* appear brighter because their far-traveling beams of light are being amplified by the effects of intense gravitational fields in the foreground.

The plane at the far left of the illustration contains background high-redshift galaxies. The Doppler shift is the apparent change in the observed frequency and wavelength of a source due to relative motion of the source and an observer. A source of light moving away from an observer experiences redshift, meaning the observed frequency is lower and the wavelength longer. Cosmological redshift is caused by the physical expansion of the universe. Astronomers estimate that the highest-redshift galaxies observed to date are about 13 billion years old.

The middle plane appearing in the illustration contains foreground galaxies. The gravity of these foreground galaxies amplifies the brightness of the background galaxies. Finally, the right plane shows how the field would look from Earth with the effects of gravitational lensing added. Distant galaxies that might otherwise be invisible appear due to gravitational lensing effects.

Einstein's general theory of relativity predicted gravitational lensing in 1915. In May 1919, the British astronomer Sir Arthur Stanley Eddington

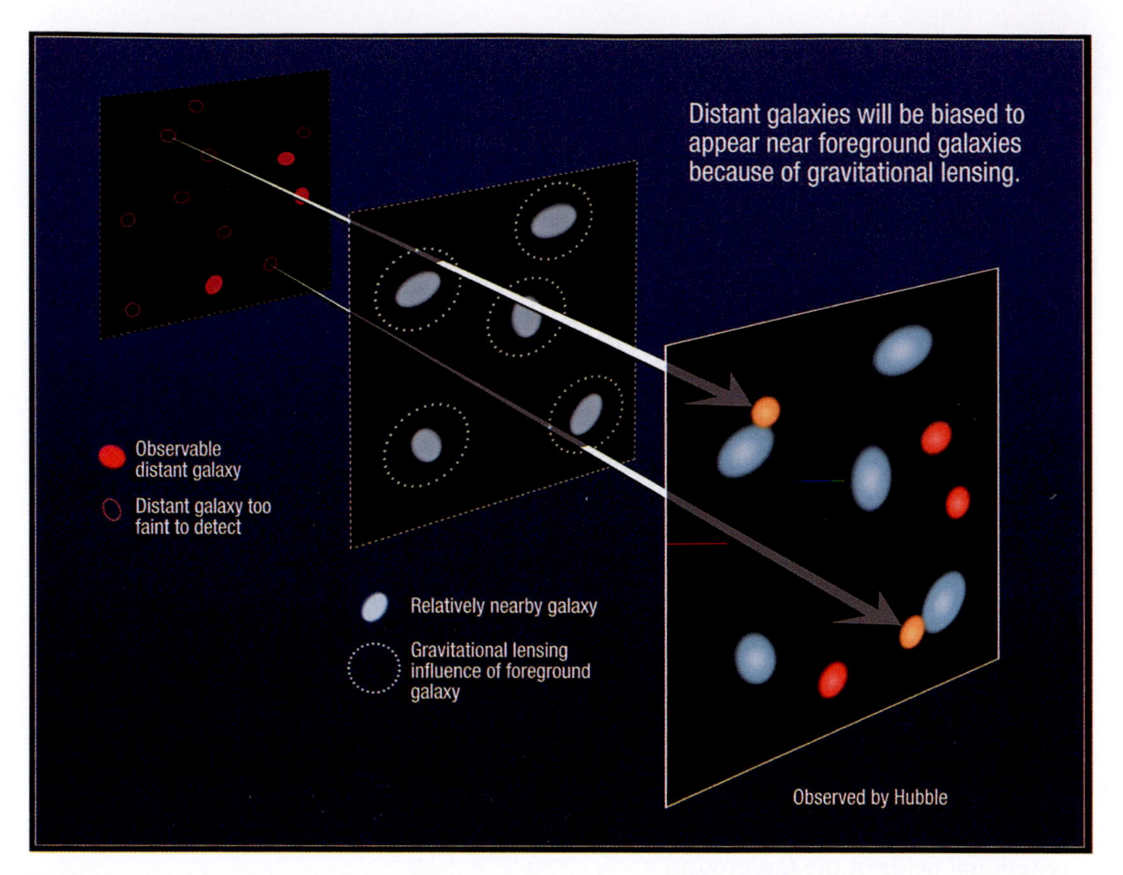

This diagram illustrates how gravitational lensing by foreground galaxies will influence the appearance of far more distant background galaxies. The plane at the far left contains background high-redshift galaxies. The middle plane contains foreground galaxies; their gravity amplifies the brightness of the background galaxies. The plane on the right shows how the field would look from Earth with the effects of gravitational lensing added. Distant galaxies that might otherwise be invisible appear due to lensing effects. *(NASA/ESA/A. Feild [STSci])*

(1882–1994) led a solar eclipse expedition to Principe Island (off the West African coast) to measure the gravitational deflection of a beam of starlight as it passed close to the Sun. Eddington's successful observation was the first confirming demonstration of Einstein's new theory of gravity and the forerunner of gravitational lensing activities by future astronomers. In the *Hubble Space Telescope* era, observational astronomers have exploited this phenomenon to find distant objects that would otherwise be invisible, including planets orbiting other stars and very distant, primeval galaxies that formed about 700 million years after the big bang.

Astronomers understand that sometimes gravitational lensing distorts a distant galaxy's appearance or alters its brightness. Other times, the gravitational lensing phenomenon splits the light from the faraway galaxy so that two or more galaxies will form around the lens, when there is really only one very distant object. When NASA's *James Webb Space Telescope* becomes operational (proposed launch date is 2014), astronomers will use its 21.3-ft. (6.5-m)-diameter primary mirror, assisted by gravitational lensing, to hunt for the most distant galaxies in the observable universe.

GRAVITATIONAL WAVES

Einstein's general theory of relativity also predicts the existence of gravity waves. The gravitational wave is gravity's analog of an electromagnetic wave. It is postulated as the phenomenon whereby gravitational radiation is emitted at the speed of light from massive compact objects experiencing rapid acceleration. Many scientists find it convenient to describe gravity waves as ripples in the fabric of space-time.

Gravitational waves should be produced by some of the following cosmic events: the binary orbit of compact massive objects such as two black holes, neutron stars, or white dwarf stars; or the merging of two galaxies. Physicists speculate that the most predictable and most powerful sources of gravitational waves emit their radiation at very low frequencies (below 10 millihertz), although some sources might be detectable at higher-frequency bands. Compact binary systems in the Milky Way galaxy are the most predictable sources of gravitational waves, while supermassive black hole merges as a result of the collision of distant galaxies should be the most powerful sources of gravitational waves.

Like water waves, gravitational waves should decrease in strength due to geometric dispersion as they propagate outward from the source. But even though they are weak, gravitational waves can travel within the fabric of space-time unobstructed by matter. Scientists have suggested constructing an extremely sensitive, space-based observatory that uses interferometry to detect gravitational waves. Scientists anticipate such detection systems will allow them to test Einstein's predictions about the existence of gravity waves and their propagation at the speed of light.

Although scientists have not yet achieved direct detection of gravitational waves, the 1993 Nobel Prize in physics was awarded to the American astrophysicists Russell Alan Hulse (1950–) and Joseph Hooton

Taylor, Jr. (1941–) for "the discovery of a new type of pulsar, a discovery that has opened up new possibilities for the study of gravitation." Their amazing discovery provided the first indirect evidence for the existence of gravitational waves.

In 1974, the Princeton University astronomers Hulse and Taylor located the pulsar called PSR 1913+16. They detected a radio pulse from PSR 1913+16 every 59 milliseconds. The pulsar orbits another compact object, which is probably another neutron star. These compact massive bodies form a binary system and orbit each other at very high speed every eight hours.

Four years after Hulse and Taylor first discovered PSR 1913+16 and after some very precise timing measurements of the pulsar, the two scientists discovered that the two stars were moving closer to each other by

This artist's rendering depicts two dense white dwarfs in the binary star system called J0806, orbiting each other once every 321 seconds. Einstein's theory of general relativity predicts that the white dwarfs in this remarkable death spiral about 1,600 light-years from Earth will lose their orbital energy by generating gravitation waves. *(NASA/GSFC/CXC [Dana Berry])*

about three millimeters per orbit. This could only happen if some phenomenon or process was removing energy from the system.

Einstein's general theory of relativity predicts that two massive objects moving around in an intense gravitational field should send gravitational waves out into space. The emission of gravitational radiation removes energy from the orbiting objects, causing them to fall gradually closer to each other. Calculations based on Einstein's general relativity suggest that the observed eight-hour orbit should get 75 microseconds shorter each year.

After 18 years of careful measurements, Taylor precisely timed the orbital periods of binary system objects in PSR 1913+16 and found that the observed orbital periods agree within 0.3 percent of the values predicted by Einstein's general theory of relativity. These results are compelling (indirect) evidence of the existence of the gravitational waves predicted by Einstein. Astronomers now estimate that it will take about 300 million years for the two neutron stars to collide.

Data from NASA's *Chandra X-ray Observatory* provided astronomers intriguing information about two white dwarf stars in a close binary orbit. Known to astronomers as J0806, this binary system is about 1,600 light-years distance from Earth. Separated by just 49,690 miles (80,000 km), the two white dwarfs orbit each other once every 321 seconds. Astrophysicists regard this cosmic death spiral (see the illustration on page 174) as one of the brightest sources of gravitational waves in the Milky Way galaxy. Einstein's general theory of relativity predicts that the two white dwarf stars should lose their orbital energy by generating gravitational waves that are directly detectable by future space-based observatories.

QUANTUM GRAVITY

Quantum gravity is the general name physicists have given to any physical theory that can successfully describe the gravitational interactions of matter and energy, when the matter and energy are being described by quantum theory. In the great majority of theories about quantum gravity, scientists insert the graviton. The graviton is a hypothetical elementary particle within quantum gravity that some physicists suggest plays a role similar to that of the photon in quantum electrodynamics (QED). Although experiments have not yet provided direct physical evidence of the existence of the graviton, Einstein predicted the existence of a quantum (or particle) of gravitational energy as early as 1915, when he was

WORMHOLES

An examination of Einstein's general theory of relativity has encouraged some scientists to speculate that matter falling into a black hole may actually survive. They suggest that under very special circumstances such matter might be conducted by means of passageways called wormholes to emerge in another place or time in this universe, or perhaps even in another universe. These hypothetical wormholes, sometimes called Einstein-Rosen bridges, are considered to be distortions or holes in the fabric of space-time. If wormholes really do exit, then in principle a person might be able to use one to travel faster than light—visiting distant parts of the universe—or possibly traveling though time as well as through space.

This is an artist's concept of a hypothetical spacecraft in the distant future traveling through a wormhole. *(NASA/GRC)*

formulating his general relativity theory. In effect, the hypothetical graviton allows the physicist to quantize gravity and model this large-scale natural force into the realm of quantum mechanics.

Einstein's general relativity theory describes gravity in terms of the curvature of space-time by matter and energy and has successfully represented the behavior of the universe on a cosmic scale. Thus, the assumption of the graviton suggests some sort of quantization of the space-time continuum itself. This shift in thinking requires some profound changes in physics and represents an enormous intellectual challenge.

So far emphasis in this chapter has been placed on two of Einstein's three great theories: special relativity and general relativity. In 1905, special relativity gave the human race the famous formula $E = \Delta mc^2$, which led not only to the first nuclear weapon but also unlocked the secrets of how stars released their energy. In 1915, general relativity provided a new perspective on the nature of gravity, as the warping or bending of the space-time continuum. His new perspective on gravitation enriched the intellectual landscape of physics with discussions such exotic objects as black holes and space warps.

Most people do not realize, however, that Einstein then spent the remainder of his professional life trying to develop a theory of everything, or TOE. Although he was unsuccessful in pursuit of the one equation that successfully explained all physical phenomena, his unified field theory would have been the greatest of his three major theoretical achievements. But even in failure, Einstein exerted a profound influence on science. His unfinished work influenced many other physicists to join the quest for a theory that neatly connects the force of gravitation to the three forces that dominate quantum mechanics (strong force, weak force, and electromagnetism).

Mathematical and theoretical advances at the end of the 20th century have revived Einstein's dream. Some form of string theory may hold the key to constructing a quantum theory of gravity, which many physicists regard as the essential step in linking gravity to the three other fundamental forces in nature. It will take future experiments at powerful accelerator facilities to provide physicists guidance as they compare theoretical predictions with real-world experimental data.

A thorough discussion of string theory (and its numerous variants) lies beyond the scope of this chapter. However, a few comments will highlight the anticipated pathways that could allow future physicists to finally complete Einstein's dream. Basically, string theory hypothesizes that the fundamental constituents of the universe are not pointlike particles but rather incredibly small, one-dimensional, vibrating strings. Each string (straight line or loop) would have its own characteristic frequency. String theory embraces and requires gravity, thereby accommodating the construction of a unified theory of all the fundamental forces and particles. However, string theory also requires the existence of six or seven extra spatial dimensions. This theoretical line of thinking has not yet been verified by any experiments. However, scientists who support string theory are optimistic that future versions of the concept will make predictions that can be experimentally tested, most likely using powerful accelerators.

The study of string theory has also led to the concept of supersymmetry (SUSY). Supersymmetry is an extension of the standard model (SM). Much like the discovery of antimatter in the 1930s, SUSY (if true) would essentially double the world of particles.

Basic supersymmetry theory suggests that the two types of particles found in the universe (bosons and *fermions*) cannot exist without each other. According to SUSY, for each type of fermion that exists, there also exists a boson (called the superpartner) with many of the same properties. In many of the theoretical variations of the supersymmetry concept, the lightest superpartner particle is stable and weakly interacting—making it a good candidate for dark matter. Although not yet verified by experiments, physicists who endorse SUSY anticipate major discoveries will be made in the next decade or so as a result of pioneering high-energy experiments performed at powerful accelerator facilities, such as CERN's Large Hadron Collider in Switzerland. This machine should have enough energy to produce all or most of the superpartner particles, either directly or through the decays of other superpartners. This line of experimentation will allow scientists to determine the pattern of superpartner masses and decays.

Supersymmetry predicts that each boson has a fermionic superpartner and that each fermion has a bosonic superpartner. These superpartners should have the same attributes as their ordinary counterparts, except for spin, which will differ by ½ unit. For example, the standard model's top quark is a fermion, spin ½ particle; its superpartner would be a boson, spin 0 particle, called the stop quark.

CHAPTER 10

Living and Thinking Matter

"Cogito ergo sum" *(I think, therefore I am.)*
—René Descartes (1596–1650), French philosopher

This chapter deals with perhaps the most extreme form of matter—the matter that makes up living organisms, especially life-forms capable of thinking and self-awareness. Mind physics, or the scientific study of consciousness, is a vibrant frontier area in contemporary physics. Closely paralleling this research field is the pursuit of machine intelligence (MI), or artificial intelligence (AI). When scientists extend the search for life beyond the boundaries of Earth, they use the fields of astrobiology and SETI. Astrobiology involves the scientific search for and study of living organisms found on celestial bodies beyond Earth. The search for extraterrestrial intelligence (SETI) is the scientific attempt to answer the following important philosophical question: Are humans alone in the universe? The chapter also involves highly speculative, though fascinating discussions about life beyond Earth.

THE BIG BANG AND LIFE

As described earlier, scientists have accumulated compelling evidence indicating that the universe began about 13.7 billion years ago as a result of an incredibly enormous explosion. Life ultimately arose out of the embers of this ancient cauldron—at least in one cosmic location, Earth.

During the universe's first few moments, an intensely energetic swarm of particles and antiparticles emerged and then promptly annihilated one another. As a result of these intensely energetic annihilation reactions, the infant universe became saturated with light (radiant energy). The very tiny fraction of the surviving normal particles (protons, neutrons,

This artist's rendering depicts a hypothetical young planet around a star that is cooler than the Sun. A soupy mixture of potentially life-forming chemicals can be seen pooling around the base of the jagged rocks. Observations from NASA's *Spitzer Space Telescope* hint that planets around cool stars—such as red dwarfs (or M-dwarfs) that are widespread throughout the Milky Way galaxy—might possess a different mixture of prebiotic (life-forming) chemicals than were present on a young planet Earth. Scientists suggest that life on Earth arose about 3.5 billion years ago out of a pond scum–like mixture of prebiotic chemicals. *(NASA/JPL-Caltech)*

PRINCIPLE OF MEDIOCRITY

The principle of mediocrity is a general assumption or speculation often used in discussions concerning the nature and probability of extraterrestrial life. It assumes that things are largely the same all over the universe—that is, it assumes that there is nothing special about Earth or humans' solar system. By invoking this hypothesis, astronomers and astrobiologists are guessing that other parts of the universe are largely as they are here. This philosophical position allows them to consider what they know about Earth—the chemical evolution of life that occurred and the facts they are discovering about other objects in this solar system—and extrapolate these data to develop concepts of what may be occurring on alien worlds around distant suns.

Today, scientists cannot pass final judgment on the validity of the principle of mediocrity. They must, at an absolute minimum, wait until human and robot explorers have made more detailed investigations of the interesting objects in humans' solar system. Celestial objects at the top of any astrobiological interest list are the planet Mars and certain moons of the giant outer planets Jupiter (particularly Europa) and Saturn (particularly Titan and Eceladus). Once scientists have explored these alien worlds in depth, they will have a much more accurate technical basis for suggesting the validity of the principle of mediocrity. The contemporary search for potentially habitable exoplanets around alien suns is also collecting data relevant to the principle of mediocrity.

and electrons) eventually became all the observable matter in this vast, expanding universe. Astrophysicists estimate that there was one extra proton for every 10 billion proton-antiproton pairs. What is especially remarkable is that the entire observable universe (all the stars, planets, and galaxies) emerged from this tiny percentage of surviving particles.

When the universe reached three minutes old, it had become an intensely hot gas composed of hydrogen nuclei, helium nuclei, and a few lithium nuclei, accompanied by a swarm of electrons. All the other chemical elements—such as carbon, oxygen, phosphorous, sulfur, and nitrogen (so necessary for life on Earth and possibly elsewhere in the universe)—did not yet exist. Without the subsequent formation of stars, the universe would have remained lifeless.

Astrophysicists now estimate that it took about 400 million years after the big bang for the first stars to form. The ancient stars were massive with very short lives, dying in spectacular supernova explosions. The life and death of these ancient stars produced the rich treasure of chemical elements beyond hydrogen and helium (see the periodic table in the appendix) and scattered these substances throughout the then lifeless universe. Giant molecular clouds of hydrogen, now enriched with the new chemical elements, condensed into new stars. During the next generation (and subsequent generations) of star formation, the presence of the new chemical elements allowed planets to form around the new stars.

At least one planet with an elementally rich solid surface, liquid water, and atmosphere formed around a new star and nourished a rich diversity of life. The American astronomer Carl Edward Sagan (1934–96) most eloquently captured the cosmic heritage of the human race with the phrase, "We are made of stardust." Today, some scientists look deep into this magnificent universe and continue to ask the profound question: Are we alone? If the miracle of life in all its biological diversity successfully journeyed down the long pathway of cosmic evolution from the big bang to 21st-century Earth, has it occurred elsewhere in a vast expanding universe filled with billions of galaxies, each of which contains billions of individual stars?

LIFE IN THE UNIVERSE

Any search for life in the universe requires that scientists develop and agree upon a basic definition of what life is. For example, according to many astrobiologists and biophysicists, life (in general) can be defined as a living system that exhibits the following three basic characteristics: (1) It is structured and contains information. (2) It is able to replicate itself. (3) It experiences few random changes in its information package—which random changes when they do occur enable the living system to evolve in a Darwinian context (that is, the survival of the fittest).

The history of life in the universe can be explored in the context of a grand, synthesizing scenario called cosmic evolution. This sweeping scenario links the development of galaxies, stars, planets, life, intelligence, and technology and then speculates on where the ever-increasing complexity of matter is leading. The emergence of complex, living matter (especially conscious intelligence), the subsequent ability of a portion of the universe to reflect upon itself, and the destiny of this intelligent consciousness are topics often associated with contemporary discussions of cosmic evolution. One interesting speculation is the anthropic principle—

namely, was the universe designed for life, especially the emergence of human life?

The cosmic evolution scenario is not without scientific basis. The occurrence of organic compounds in interstellar clouds, in the atmospheres of the giant planets of the outer solar system, and in comets and meteorites suggests the existence of a chain of astrophysical processes that links the chemistry of interstellar clouds with the prebiotic evolution of organic matter in the solar system and on early Earth. There is also compelling evidence that cellular life existed on Earth some 3.56 billion years ago (3.56 Gy). This implies that the cellular ancestors of contemporary terrestrial life emerged rather quickly (on a geologic timescale). These ancient creatures may have also survived the effects of large impacts from comets and asteroids in those ancient, chaotic times when the solar system was evolving.

ANTHROPIC PRINCIPLE

This interesting, though highly controversial, hypothesis in modern cosmology suggests that the universe evolved after the big bang in just the right way so that life, especially intelligent life, could develop. The proponents of this hypothesis contend that the fundamental physical constants of the universe actually support the existence of life and (eventually) the emergence of conscious intelligence—including, of course, human beings. The advocates of this hypothesis further suggest that with just a slight change in the value of any of these fundamental physical constants, the universe would have evolved very differently after the big bang.

If the force of gravitation were weaker than it is, expansion of matter after the big bang would have been much more rapid, and the development of stars, planets, and galaxies from extremely sparse (nonaccreting) nebular materials might not have occurred. No stars, no planets, no development of life as scientists currently know it. If, on the other hand, the force of gravitation were stronger than it is, the expansion of primordial material would have been sluggish and retarded, encouraging a total gravitational collapse (that is, the big crunch) long before the development of stars and planets.

Opponents of this hypothesis suggest that the values of the fundamental physical constants are just a coincidence. Until this hypothesis can actually be tested within the rigors of the scientific method, it must remain out of the mainstream of demonstrable science.

Scientists have identified several factors they consider important in the evolution of complex life. These include (a) endogenous factors stemming from physical-chemical properties of Earth, and those of eukaryotic organisms; (b) factors associated with properties of the Sun and of Earth's position with respect to the Sun; (c) factors originating within the solar system, including Earth as a representative planet; and (d) factors originating in space far from humans' solar system.

The word *eukaryotic* refers to cells whose internal construction is complex, consisting of organelles (e.g., nucleus, mitochondria, etc.), chromosomes, and other structures. All higher terrestrial organisms are built of eukaryotic cells, as are many single-celled organisms (called protists). The evolution of complex life apparently had to await the evolution of eukaryotic cells—an event that is believed to have occurred on Earth about 1 billion years ago. A eukaryote is an organism built of eukaryotic cells.

All living things are extremely complex, interesting collections of matter. Life-forms that have achieved intelligence and have developed technology are especially valuable in the cosmic evolution of the universe. Intelligent creatures with technology, including human beings on Earth, can exercise conscious control over matter in progressively more effective ways as the level of their technology grows. Ancient cave dwellers used fire to provide light and warmth. Modern humans harness solar energy, control falling water, and split atomic nuclei to provide energy for light, warmth, industry, and entertainment. Later in this century, their children or grandchildren will most likely join atomic nuclei (controlled fusion) to provide energy for light, warmth, industrial applications, and entertainment on Earth, as well as for interplanetary power and propulsion systems as the human race emerges into the solar system with settlements on the Moon and Mars.

Some scientists even speculate that if technologically advanced civilizations throughout the Milky Way galaxy can learn to live with the awesome powers unleashed in such advanced technologies, then it may be the overall destiny of advanced intelligent life-forms, including human beings, to ultimately exercise (beneficial) control over all the matter and energy in their portion of the universe.

According to modern scientific theory, living organisms arose naturally on the primitive Earth through a lengthy process of chemical evolution of organic matter. This process began with the synthesis of simple organic compounds from inorganic precursors in the atmosphere; continued in the oceans, where these compounds were transformed into increasingly more complex organic substances; then culminated with the emergence of organic microstructures that had the capability of rudimentary self-replication and other biochemical functions.

The essentially universal presence of these compounds throughout interstellar space gives astrobiologists the scientific basis for forming the following important contemporary hypothesis: The origin of life is inevitable throughout the universe, wherever these compounds occur and suitable planetary conditions exist. Present-day understanding of life on Earth leads modern scientists to the conclusion that life originates on planets and that the overall process of biological evolution is subject to the often chaotic processes associated with planetary and solar system evolution—for example, the random impact of a comet on a planetary body or the unpredictable breakup of a small moon.

The continuing discovery of exoplanets and revisions in planetary formation theory now strongly suggest that objects similar in mass and composition to Earth probably exist in many planetary systems. In order to ascertain whether any of these extrasolar planets are capable of sustaining life (as scientists understand it on Earth), astrobiologists need to use a family of advanced planet-hunting systems, such as NASA's *Kepler* spacecraft (launched in 2009), to accomplish direct detection of terrestrial (that is, small and rocky) planetary companions to other stars, as well as to investigate the composition of their atmospheres. Liquid water is a basic requirement for life as we know it, and it is the key indicator that will be used by scientists to determine whether planets revolving around other stars may indeed be life sustaining.

CONSCIOUSNESS AND THE UNIVERSE

The history of the universe can also be viewed to follow a more or less linear timescale. The scenario of cosmic evolution links the development of the galaxies, stars, heavy elements, life, intelligence, technology, and the future. It is especially useful in philosophical and theological cosmology.

Astrobiologists are interested in understanding how life, especially intelligent life and consciousness, can emerge out of the primordial matter from which the galaxies, stars, and planets evolved. This approach leads to such interesting cosmological concepts as the living universe, the conscious universe, and the thinking universe. Is the evolution of intelligence and consciousness a normal end point for the development of living matter everywhere in the universe? Or are human beings unique

EXOTHEOLOGY

Exotheology is the organized body of opinions concerning the impact that space exploration and the possible discovery of life beyond the boundaries of Earth would have on contemporary terrestrial religions. On Earth, theology involves the study of the nature of God and the relationship of human beings to God.

Throughout human history, people have gazed into the night sky and pondered the nature of God. They searched for those basic religious truths and moral beliefs that define how an individual should interact with his or her Creator and toward one another. The theologies found within certain societies, especially ancient ones, sometimes involved a collection of gods (polytheism). For people in such societies, the plurality of specialized gods proved useful in explaining natural phenomena, as well as in codifying human behavior. For example, in the polytheism of ancient Greece, an act that really displeased Zeus—the lord of the gods who ruled Earth from Mount Olympus—would cause him to hurl a powerful thunderbolt at the offending human. The fear of getting "zapped" by Zeus may have helped ancient Greece societies maintain a set of moral standards. Other terrestrial religions stressed belief in a single God. Judaism, Christianity, and Islam represent planet Earth's three major monotheistic religions. Each has a collection of dogma and beliefs that define acceptable moral behavior by human beings and also identify individual rewards (for example, personal salvation) and punishments (for example, eternal damnation) for adherence or transgressions, respectively.

by-products of the cosmic evolution process—perhaps the best the universe could do since the big bang?

If the answer to the former question is positive, then the universe should be teeming with life, including intelligent life. If the latter answer is possible, human beings could be alone in a very big universe. Furthermore, if humans are the only beings now capable of contemplating the universe, then perhaps it is also the destiny of the human race to venture to the stars and carry life and consciousness to places where now there is only the material potential for such.

Up until recently, most scientists have avoided integrating the potential role of conscious intelligence (i.e., matter that can think) in cosmological models. But what happens after conditions in the universe

In this century, advanced space exploration missions could lead to the discovery of simple alien life-forms on other worlds within this solar system. The important scientific achievement would undoubtedly rekindle serious interest in one of the oldest philosophical questions that has puzzled a great number of people throughout history: Are humans the only intelligent species in the universe? If not, what happens when contact is made with an alien intelligence? At this point, it is difficult to anticipate the full effect interaction with intelligent alien creatures (should they exist) would have on terrestrial philosophies and religions.

Some space-age theologians are beginning to grapple with these intriguing questions and many similar ones. For example, should it exist, would an alien civilization that is much older than human planetary civilization have a significantly better understanding of the universe and, by extrapolation, a clearer understanding of the nature of God? On Earth, many brilliant scientists, such as Sir Isaac Newton and Albert Einstein, regarded their deeper personal understanding of the physical universe as an expanded perception of its Creator. Would these (hypothetical) advanced alien creatures be willing to share their deeper theological insights? And if they do decide to share such insights into the divine nature, what effect would that communication have on terrestrial religions? Exotheology involves such interesting speculations and the extrapolation of teachings found within terrestrial religions to accommodate the expanded understanding of the universe brought about by the scientific discoveries of space exploration and modern astronomy.

give rise to living matter that can think and reflect upon itself and its surroundings? The anthropic principle is an interesting, though highly controversial, hypothesis in contemporary cosmology. As previously mentioned, it suggests that the universe has evolved after the big bang in just the right way so that life, especially intelligent life, could develop on Earth. Does human intelligence, or possibly other forms of alien intelligence, have a role in the further evolution and destiny of the universe? If scientists and civilian authorities thought the Copernican hypothesis livened up intellectual activities in the 16th and 17th centuries, the bold hypothesis set forth in the anthropic principle should keep scientists, philosophers, and theologians very busy for a good portion of this century.

MACHINES THAT THINK

Perhaps the most intriguing philosophical question involved with advanced robot systems is the question of machine intelligence and machine consciousness. Simply stated: Is a machine that thinks conscious and aware of its existence? The French philosopher René Descartes (1596–1650) believed that the bodies of humans and animals are complex automata. In his treatise *Discourse on Method,* published in 1637, Descartes discusses how humans, who have the power of reason, and animals, which cannot reason, can be distinguished from one another and machines. In *Discourse on Method,* he famously states: *"Cogito ergo sum,"* which means, "I think, therefore I am." This statement highlights some of the deep philosophical arguments Descartes raised in developing his mind-body dualism.

The nature of consciousness and the human mind is an issue that has intrigued philosophers, physicians, and scientists for centuries. Interest in this issue peaked again when robot specialists started speculating about endowing very smart machines with a sense of consciousness and cognition. At what point does a silicon-based, thinking machine become truly conscious and comparable (in thought capacity) with carbon-based, human intelligence?

In 1950, the British mathematician and computer science pioneer Alan Mathison Turing (1912–54) raised a similar question in his intriguing paper "Computing machines and intelligence." As part of his pioneering discussion on artificial intelligence, Turing gave the world a test, now called the Turing test, for judging whether a machine is successfully simulating the thought processes of the human mind.

Artificial intelligence (AI) is a term commonly taken to mean the study of thinking and perceiving as general information-processing functions—or the science of machine intelligence (MI). Computer systems have been programmed to diagnose diseases; prove theorems; analyze electronic circuits; play complex games such as chess, poker, and backgammon; solve differential equations; assemble mechanical equipment, using robotic manipulator arms and end effectors (the "hands" at the end of the manipulator arms); guide robotic (uncrewed) vehicles across complex terrestrial terrain, as well as through the vast reaches of interplanetary space; analyze the structure of complex organic molecules; understand human speech patterns; and write other computer programs. All of these computer-accomplished functions required a degree of intelligence similar to mental activities performed by the human brain.

SELF-REPLICATING SYSTEM (SRS)

The self-replicating system is a postulated type of very advanced thinking-and-doing machine. A single SRS unit is a machine system that contains all the elements required to maintain itself, to manufacture desired products, and even, as the name implies, to reproduce itself.

John von Neumann was the first person to seriously consider the problem of self-replicating machine systems. During and following World War II, he became

(continues)

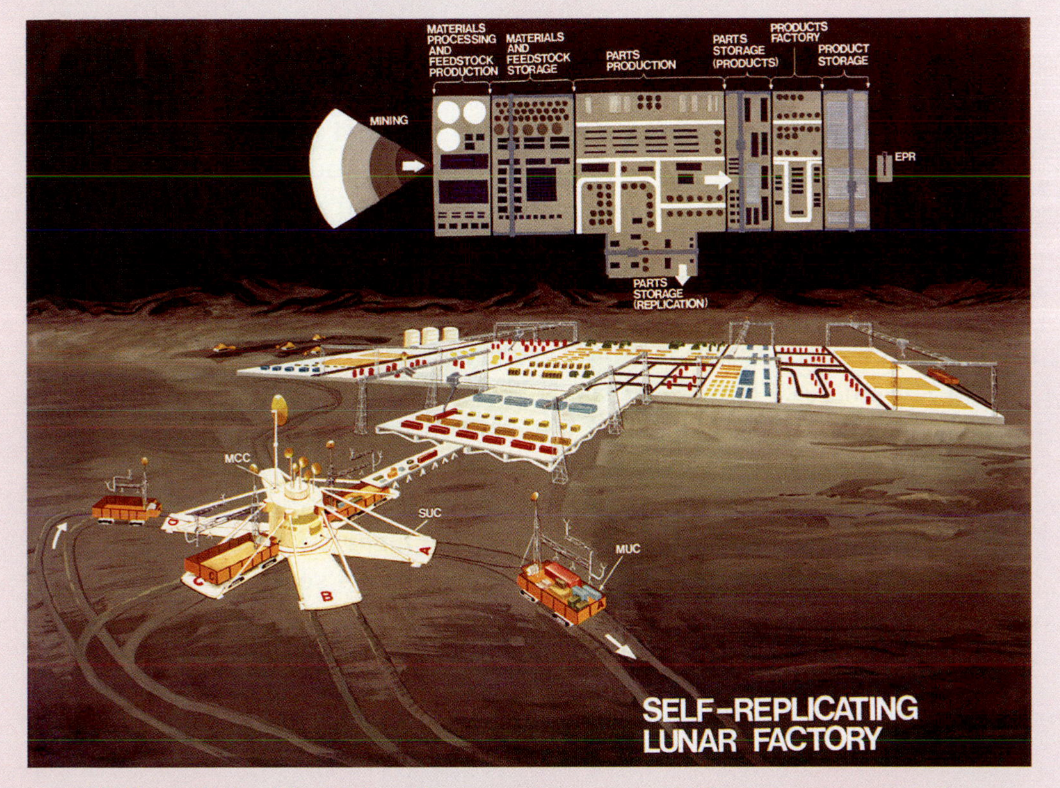

An artist's rendering illustrating the general structure and basic components of a conceptual self-replicating lunar factory. This SRS unit has four major subsystems. The materials-processing subsystem gathers raw materials from the extraterrestrial environment (here the lunar surface) and prepares industrial feedstock. Next, a parts-production subsystem uses the feedstock to manufacture other parts or entire new machines. The factory has two basic outputs: a new self-replicating factory or commercially useful products. New SRS units are produced until a preselected SRS unit population is reached; at that point, each factory begins producing commercial products. ***(NASA)***

(continued)

interested in the study of automatic replication as part of his wide-ranging interests in complicated machines. From von Neumann's initial work and the more recent work of other investigators, five general classes of SRS behavior have been defined: production, replication, growth, repair, and evolution.

The issue of closure (total self-sufficiency) is one of the fundamental problems in designing self-replicating systems. In an arbitrary SRS unit, there are three basic requirements necessary to achieve total closure: (1) matter closure, (2) energy closure, and (3) information closure. If the machine device is only partially self-replicating, then it is said that only partial closure of the system has occurred. In this case, some essential matter, energy, or information must be provided from external sources, or else the machine would fail to reproduce itself.

The self-replicating system could be used to assist human beings in planetary engineering projects, where a seed machine would be sent to a target celestial body, such as the Moon or Mars. The SRS unit would make a sufficient number of identical copies (using extraterrestrial resources) and then set about some production task, such as manufacturing oxygen to provide a breathable atmosphere for future human settlers.

The extreme extrapolation of this scenario involves human beings sending a seed SRS unit into interstellar space as part of a long-term plan to trigger a wave of galactic exploration. Over the ensuing millennia, the initial SRS unit would stop in the various solar systems it encounters on its cosmic journey to perform repairs and to make copies of itself. If suitable exoplanets are encountered, the SRS might also execute protocols that plant the seeds of life.

An enormously large self-replicating system could serve as an interstellar space ark that allows human beings to initiate a wave of carbon-based life (as understood on Earth) and machine intelligence throughout the galaxy. How far this wave of life and consciousness would successfully propagate across the Milky Way is anyone's guess at this point.

Sometime later in this century, a general theory of intelligence should emerge from the current efforts of AI scientists and engineers. This general theory would help to guide the design and development of much smarter robot spacecraft and exploratory probes, allowing humans to explore more fully the solar system and use its material resources.

Artificial intelligence generally includes a number of elements, or subdisciplines: planning and problem solving; perception; natural language; expert systems; automation, teleoperation, and robotics; distributed data management; and cognition and learning.

Automatic devices are those that operate without direct human control. NASA has already used many highly automated, smart machines to explore alien worlds. Teleoperation implies that a human operator is in remote control of a mechanical system. Control signals can be sent by means of hardwire (if the device under control is nearby) or via electromagnetic signals (for example, laser or radio frequency), if the robot system is some distance away. Starting in 1997, NASA's missions to the surface of Mars began demonstrating teleoperation of robot rover spacecraft across planetary distances. The robot rover vehicles were teleoperated by the Earth-based flight team at the Jet Propulsion Laboratory (JPL). The human operators used images of the Martian surface obtained by both the rover and the lander systems. These interplanetary teleoperations required that the rover be capable of some semi-autonomous operation since there was a time delay of the signals that averaged between 10 and 15 minutes duration—depending on the relative position of Earth and Mars over the course of the mission. Since early 2004, NASA's *Spirit* and *Opportunity* surface rovers provided even more sophisticated teleoperation experiences at interplanetary distances.

In dealing with the great distances in interplanetary exploration, a situation is eventually reached when electromagnetic wave transmission cannot accommodate effective near real-time control. When the device to be controlled on an alien world is a few light-hours away and when actions or discoveries require split-second decisions, teleoperation must yield to increasing levels of autonomous, machine intelligence-dependent robotic operation. For example, when it encounters Pluto in July 2015, NASA's *New Horizons* spacecraft will be more than 3 billion miles (4.8 billion km) from Earth. At such enormous distances, the spacecraft is essentially on its own with respect to executing all the protocols of the scientific flyby of the icy dwarf planet.

Robot devices are computer-controlled mechanical systems that are capable of manipulating or controlling other machine devices, such as end effectors. Robots may be mobile or fixed in place and either fully automatic or teleoperated. Large quantities of data are frequently involved in the operation of automatic robotic devices. The field of distributed data management is concerned with ways of organizing cooperation among independent but mutually interacting databases.

In the field of artificial intelligence, the concept of cognition and learning refers to the development of a machine intelligence that can deal with new facts, unexpected events, and even contradictory information. Today's smart machines handle new data by means of preprogrammed methods or logical steps. Tomorrow's smarter machines will need the ability to learn, possibly even to understand, as they encounter new situations and are forced to change their mode of operation.

Perhaps late in this century, as the field of artificial intelligence sufficiently matures, people will send fully automatic robot probes on interstellar voyages. Each very smart interstellar probe must be capable of independently examining a new star system for suitable (life-potential) exoplanets and, if successful in locating one, beginning the search for extraterrestrial life. Meanwhile, back on Earth, scientists will wait for the probe's electromagnetic signals to travel light-years across the interstellar void, eventually informing its human builders that the extraterrestrial exploration plan has been successfully accomplished.

In his posthumously published book *Theory of Self-Replicating Automata,* the brilliant Hungarian-born, German-American mathematician John von Neumann (1903–57), shared some of his visionary ideas about truly advanced robots (automata), which would be capable of making copies of themselves and performing all manner of construction tasks. If ever created, a self-replicating system (SRS)—sometimes referred to as a von Neumann machine—would have a profound impact on how human beings manipulate and control energy and matter resources on Earth, throughout the solar system, and beyond.

CHARACTERISTICS OF EXTRATERRESTRIAL CIVILIZATIONS

This final section contains some very interesting, though highly speculative, discussions concerning the characteristics of intelligent alien civilizations—should they exist in the Milky Way galaxy.

According to some scientists, intelligent life in the universe might be thought of as experiencing three basic levels of civilization. In 1964, the Russian astronomer Nikolai Semenovich Kardashev (1932–), while examining the issue of information transmission by extraterrestrial civilizations, postulated three types of technologically developed civilizations on the basis of their energy use. The Kardashev Type I civilization would represent a planetary civilization similar to the technology level on

Earth today. It would command the use of somewhere between 10^{12} and 10^{16} watts of energy—the upper limit being the amount of solar energy being intercepted by a suitable (or Goldilocks) planet in its orbit about the parent star.

Using humans' solar system as a reference, a Kardashev Type II civilization could ultimately command between 10^{26} and 10^{27} watts (J/s) of radiant energy. This hypothetical civilization might engage in feats of planetary engineering, emerging from its native planet through advances in space technology and extending its resource base throughout the local star system. The eventual upper limit of a Type II civilization could be taken as the creation of a Dyson sphere. A Dyson sphere is a postulated shell-like cluster of habitats and structures placed entirely around a star by an advanced civilization to intercept and basically use all the radiant energy from that parent star. What the British-American physicist Freeman John Dyson (1923–) suggested in 1960 was that an advanced extraterrestrial civilization might eventually develop the space technologies necessary to rearrange the raw materials of all the planets in its solar system, creating a more efficient composite ecosphere around the parent star. Dyson further hypothesized that any such advanced alien civilizations might be detected by the presence of thermal infrared emissions from these artificially enclosed star systems in contrast to the normally anticipated visible radiation. Once this level of extraterrestrial civilization is achieved, the search for additional resources and the pressures of continued growth could encourage interstellar migrations. This would mark the start of a Kardashev Type III extraterrestrial civilization.

At maturity, the Kardashev Type III civilization would be capable of harnessing the material and energy resources of an entire galaxy (typically containing some 10^{11} to 10^{12} stars). Energy resources on the order of 10^{37} to 10^{38} watts or more would be involved.

Command of energy resources might therefore represent a key factor in the evolution of extraterrestrial civilizations. It should be noted that a Type II civilization controls about 10^{12} times the energy resources of a Type I civilization; and a Type III civilization controls approximately 10^{12} times as much energy as a Type II civilization.

Starting with Earth as a model (the one and only scientific data point now available), some scientists postulate that a Type I civilization would probably exhibit the following characteristics: (1) an understanding of the laws of physics; (2) a planetary society, including a global communication network and interwoven food and materials resource networks;

This artist's rendering provides the cockpit view of a hypothetical interstellar spacecraft traveling at 80 percent of the speed of light and the visual distortions that the crew would experience at such high speeds. The star field is actually being wrapped toward the front of the craft in addition to being significantly blueshifted. *(NASA sponsored digital art by Les Bossinas [Cortez III Service Corp.], 1998)*

(3) intentional or unintentional emission of electromagnetic radiations (especially radio frequency); (4) the development of space technology and rocket propulsion-based interplanetary space travel—critical tools necessary to leave the home planet; (5) (possibly) the development of nuclear energy technology, both power supplies and weapons; and (6) (possibly) a desire to search for and communicate with other intelligent life-forms in the universe. Of course, many uncertainties are present in such characterizations. For example, given the development of the space technology, will an alien planetary civilization decide to create a solar system civilization? Do these intelligent creatures, should they exist, develop the long-range societal-planning perspective that supports the eventual creation of artificial habitats and structures throughout their star system? Or do the majority of Type I civilizations unfortunately destroy themselves with their own advanced technologies before they can emerge from a politi-

cally unstable planetary civilization into a more socially stable Type II civilization? Does the exploration imperative encourage such creatures to go out from their comfortable planetary niche into an initially hostile but resource-rich star system? If this "cosmic birthing" does not occur frequently, perhaps the Milky Way galaxy is indeed populated with intelligent life but at a level of stagnant planetary (Type I) civilizations that have neither the technology nor the motivation to create an extraterrestrial civilization or even to try to communicate with any other intelligent life-forms across interstellar distances.

In all likelihood, the Milky Way galaxy does not presently contain a Type III civilization. Or else humans' solar system is being ignored—that is, intentionally being kept isolated—perhaps as a game preserve or zoo, as some scientists have speculated. Then again, this solar system may be simply one of the very last regions to be filled in.

Exciting as these speculations may appear, there is also another perspective. If humans are indeed alone, or the most technically advanced creatures in the galaxy, then people now stand at the technological threshold of creating the galaxy's first Type II civilization. Should the people succeed in that task, then the human race would have the opportunity of becoming the first interstellar civilization to sweep across interstellar space, founding a Type III civilization across the Milky Way galaxy.

CHAPTER 11

Conclusion

"They shall beat their swords into plowshares and their spears into pruning hooks.
Nation will not take up sword against nation, nor will they train for war anymore."

—The Prophet Isaiah (2:4)

The rate at which science has revealed the mysteries of matter and energy since the start of the Scientific Revolution is nothing short of amazing. The brief examination of the extreme states of matter that has taken place in this volume suggests that the universe is full of exciting, magical things and phenomena, which patiently wait for human minds to become more knowledgeable and discover them.

It is both inspiring and rather humbling to examine how humans started recognizing the true complexity and beauty of the universe, which emerged about 13.7 billion years ago, following a gigantic explosion called the big bang. Less than five centuries ago, human beings accepted, albeit reluctantly at first, that Earth was neither the center of the solar system nor the center of the universe. Over the next three centuries, astronomers showed that humans' parent star, the Sun, was just a rather average star among very many in the Milky Way galaxy. In the early 20th century, astronomers and astrophysicists demonstrated through careful observations that the Milky Way galaxy was itself but one of billions and billions of galaxies.

By carefully scrutinizing the cosmic microwave background, the fossilized embers of the big bang, scientists are finding unanticipated new information about how the universe formed and what it is actually made of. They discovered two incredibly important new phenomena: dark matter and dark energy. Understanding these mysterious aspects of the universe will shape and drive science for centuries.

Elaborate planetary science investigations further revealed that the solar system formed about 5 billion years ago—making the Sun and Earth less about one-third the age of the universe. By studying the energetic and violent demise of stars, including those similar to the Sun, astronomers now know that Earth will more than likely disappear when the Sun leaves the main sequence and becomes a red giant in about 5 billion years from now.

Although Earth's physical position in the universe has dramatically receded from the central (geocentric) model promoted by ancient Greek astronomers, human beings have discovered and continue to discover physical laws and relationships between matter and energy that appear to hold throughout the observable universe and will continue to do so far into the future. Unless other parts of the universe obey different physical laws (there is no evidence to support this hypothesis), each physical law, equation of state, or new natural phenomenon that scientists uncover represents an enlargement of humanity's intellectual gateway into the unimaginably intriguing cosmic realm that awaits future generations beyond the boundaries of Earth.

From a planetary perspective, Earth is a spaceship traveling through the interstellar void and all crew members must learn how to share the common supplies of air and other natural resources. This sobering thought emphasizes the need for responsible stewardship of the planet. Should the human family continue to allow reckless episodes of pollution, resource consumption, and conflict, people must brace themselves for the adverse, planetary-scale consequences of such foolhardy and self-destructive behavior.

The universe is a naturally extreme and often violent place. Everyone alive today and every person who has ever lived upon Earth shares at least one common characteristic—they are the product of atoms forged under extreme conditions. Specifically, the atoms of life found on Earth (carbon, hydrogen, oxygen, nitrogen, sulfur, and phosphorous) trace their heritage back either directly to the big bang (hydrogen), or else to the fiery interiors and catastrophic explosions of ancient stars.

About 5 billion years ago, this solar system evolved from a primordial cloud of hydrogen gas; now that same incredibly interesting gas holds the key to the future of the human race. Control of hydrogen, nature's most abundant element, represents a technical pathway to humankind's future on this planet and beyond. Hydrogen in fuel cells and as the combustible fuel for traditional thermodynamic power conversion offers a more environmentally friendly approach to electric power generation and transportation. Harnessing controlled thermonuclear fusion promises incredible possibilities here on Earth and would make the solar system available to more intense exploration and even human habitation.

Finally, the search for living matter (that is, life) beyond Earth could produce many exciting scientific discoveries that would have enormous consequences for the human race. If life and consciousness proves to be extremely rare, then future generations of human beings should recognize that they have a serious obligation to the entire (still unconscious) universe to preserve the precious biological heritage that has taken about 4 billion or so years to emerge and evolve on planet Earth. If, on the other hand, life (especially intelligent life) is found to be abundant throughout the Milky Way galaxy, then future human generations might eagerly seek to communicate with other conscious beings and discover how matter has also evolved to extreme states of complexity on alien worlds.

One important fact is clear right now: The people of Earth are children of the stars. An increased knowledge of the extreme states of matter will give future generations of human beings the exciting opportunity to make the stars both a destination and a destiny.

Appendix

Scientists correlate the properties of the elements portrayed in the periodic table with their electron configurations. Since, in a neutral atom, the number of electrons equals the number of protons, they arrange the elements in order of their increasing atomic number (Z). The modern periodic table has seven horizontal rows (called periods) and 18 vertical columns (called groups). The properties of the elements in a particular row vary across it, providing the concept of periodicity.

There are several versions of the periodic table used in modern science. The International Union of Pure and Applied Chemistry (IUPAC) recommends labeling the vertical columns from 1 to 18, starting with hydrogen (H) as the top of group 1 and ending with helium (He) as the top of group 18. The IUPAC further recommends labeling the periods (rows) from 1 to 7. Hydrogen (H) and helium (He) are the only two elements

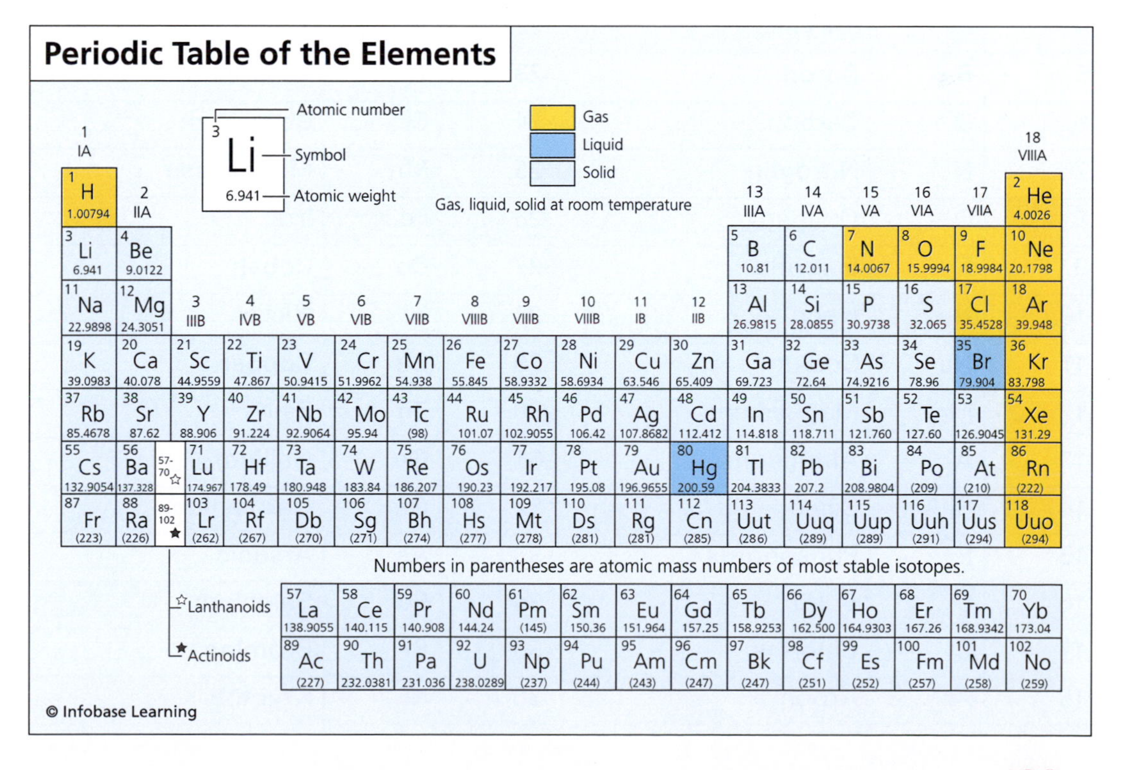

found in period (row) 1. Period 7 starts with francium (Fr) and includes the actinide series as well as the transactinides (very short-lived, human-made, super-heavy elements).

The row (or period) in which an element appears in the periodic table tells scientists how many electron shells an atom of that particular element possesses. The column (or group) lets scientists know how many electrons to expect in an element's outermost electron shell. Scientists call an electron residing in an atom's outermost shell a valence electron. Chemists have learned that it is these valence electrons that determine the chemistry of a particular element. The periodic table is structured such that all the elements in the same column (group) have the same number of valence electrons. The elements that appear in a particular column (group) display similar chemistry.

ELEMENTS LISTED BY ATOMIC NUMBER

1	H	Hydrogen	19	K	Potassium
2	He	Helium	20	Ca	Calcium
3	Li	Lithium	21	Sc	Scandium
4	Be	Beryllium	22	Ti	Titanium
5	B	Boron	23	V	Vanadium
6	C	Carbon	24	Cr	Chromium
7	N	Nitrogen	25	Mn	Manganese
8	O	Oxygen	26	Fe	Iron
9	F	Fluorine	27	Co	Cobalt
10	Ne	Neon	28	Ni	Nickel
11	Na	Sodium	29	Cu	Copper
12	Mg	Magnesium	30	Zn	Zinc
13	Al	Aluminum	31	Ga	Gallium
14	Si	Silicon	32	Ge	Germanium
15	P	Phosphorus	33	As	Arsenic
16	S	Sulfur	34	Se	Selenium
17	Cl	Chlorine	35	Br	Bromine
18	Ar	Argon	36	Kr	Krypton

37	Rb	Rubidium
38	Sr	Strontium
39	Y	Yttrium
40	Zr	Zirconium
41	Nb	Niobium
42	Mo	Molybdenum
43	Tc	Technetium
44	Ru	Ruthenium
45	Rh	Rhodium
46	Pd	Palladium
47	Ag	Silver
48	Cd	Cadmium
49	In	Indium
50	Sn	Tin
51	Sb	Antimony
52	Te	Tellurium
53	I	Iodine
54	Xe	Xenon
55	Cs	Cesium
56	Ba	Barium
57	La	Lanthanum
58	Ce	Cerium
59	Pr	Praseodymium
60	Nd	Neodymium
61	Pm	Promethium
62	Sm	Samarium
63	Eu	Europium
64	Gd	Gadolinium
65	Tb	Terbium
66	Dy	Dysprosium
67	Ho	Holmium
68	Er	Erbium
69	Tm	Thulium
70	Yb	Ytterbium
71	Lu	Lutetium
72	Hf	Hafnium
73	Ta	Tantalum
74	W	Tungsten
75	Re	Rhenium
76	Os	Osmium
77	Ir	Iridium
78	Pt	Platinum
79	Au	Gold
80	Hg	Mercury
81	Tl	Thallium
82	Pb	Lead
83	Bi	Bismuth
84	Po	Polonium
85	At	Astatine
86	Rn	Radon
87	Fr	Francium
88	Ra	Radium
89	Ac	Actinium
90	Th	Thorium
91	Pa	Protactinium
92	U	Uranium
93	Np	Neptunium
94	Pu	Plutonium

(continues)

ELEMENTS LISTED BY ATOMIC NUMBER *(continued)*

95	Am	Americium	107	Bh	Bohrium
96	Cm	Curium	108	Hs	Hassium
97	Bk	Berkelium	109	Mt	Meitnerium
98	Cf	Californium	110	Ds	Darmstadtium
99	Es	Einsteinium	111	Rg	Roentgenium
100	Fm	Fermium	112	Cn	Copernicum
101	Md	Mendelevium	113	Uut	Ununtrium
102	No	Nobelium	114	Uuq	Ununquadium
103	Lr	Lawrencium	115	Uup	Ununpentium
104	Rf	Rutherfordium	116	Uuh	Ununhexium
105	Db	Dubnium	117	Uus	Ununseptium
106	Sg	Seaborgium	118	Uuo	Ununoctium

Chronology

Civilization is essentially the story of the human mind understanding and gaining control over matter. The chronology presents some of the major milestones, scientific breakthroughs, and technical developments that formed the modern understanding of matter. Note that dates prior to 1543 are approximate.

13.7 BILLION YEARS AGO ..... Big bang event starts the universe.

13.3 BILLION YEARS AGO ..... The first stars form and begin to shine intensely.

4.5 BILLION YEARS AGO ...... Earth forms within the primordial solar nebula.

3.6 BILLION YEARS AGO ...... Life (simple microorganisms) appears in Earth's oceans.

2,000,000–100,000 B.C.E. .. Early hunters of the Lower Paleolithic learn to use simple stone tools, such as handheld axes.

100,000–40,000 B.C.E. .... Neanderthal man of Middle Paleolithic lives in caves, controls fire, and uses improved stone tools for hunting.

40,000–10,000 B.C.E. ..... During the Upper Paleolithic, Cro-Magnon man displaces Neanderthal man. Cro-Magnon people develop more organized hunting and fishing activities using improved stone tools and weapons.

8000–3500 B.C.E. .... Neolithic Revolution takes place in the ancient Middle East as people shift their dependence for subsistence from hunting and gathering to crop cultivation and animal domestication.

3500–1200 B.C.E. .... Bronze Age occurs in the ancient Middle East, when metalworking artisans start using bronze (a copper and tin alloy) to make weapons and tools.

1200–600 B.C.E. .... People in the ancient Middle East enter the Iron Age. Eventually, the best weapons and tools are made of steel, an alloy of iron and varying amounts of carbon. The improved metal tools and weapons spread to Greece and later to Rome.

1000 B.C.E. .... By this time, people in various ancient civilizations have discovered and are using the following chemical elements (in alphabetical order): carbon (C), copper (Cu), gold (Au), iron (Fe), lead (Pb), mercury (Hg), silver (Ag), sulfur (S), tin (Sn), and zinc (Zn).

650 B.C.E. .... Kingdom of Lydia introduces officially minted gold and silver coins.

600 B.C.E. .... Early Greek philosopher Thales of Miletus postulates that all substances come from water and would eventually turn back into water.

450 B.C.E. .... Greek philosopher Empedocles proposes that all matter is made up of four basic elements (earth, air, water, and fire) that periodically combine and separate under the influence of two opposing forces (love and strife).

430 B.C.E. .... Greek philosopher Democritus proposes that all things consist of changeless, indivisible, tiny pieces of matter called *atoms*.

250 B.C.E. .... Archimedes of Syracuse designs an endless screw, later called the Archimedes screw. People use the fluid-moving device to remove water from the holds of sailing ships and to irrigate arid fields.

300 C.E. .... Greek alchemist Zosimos of Panoplis writes the oldest known work describing alchemy.

850 .... The Chinese use gunpowder for festive fireworks. It is a mixture of sulfur (S), charcoal (C), and potassium nitrate (KNO_3).

1247 .... British monk Roger Bacon writes the formula for gunpowder in his encyclopedic work *Opus Majus*.

1250 .... German theologian and natural philosopher Albertus Magnus isolates the element arsenic (As).

1439 .... Johannes Gutenberg successfully incorporates movable metal type in his mechanical printing press. His revolutionary approach to printing depends on a durable, hard metal alloy called type metal, which consists of a mixture of lead (Pb), tin (Sn), and antimony (Sb).

1543 .... Start of the Scientific Revolution. Polish astronomer Nicholas Copernicus promotes heliocentric (Sun-centered) cosmology with his deathbed publication of *On the Revolutions of Celestial Orbs.*

1638 .... Italian scientist Galileo Galilei publishes extensive work on solid mechanics, including uniform acceleration, free fall, and projectile motion.

1643 .... Italian physicist Evangelista Torricelli designs the first mercury barometer and then records the daily variation of atmospheric pressure.

1661 .... Irish-British scientist Robert Boyle publishes *The Sceptical Chymist,* in which he abandons the four classical Greek elements (earth, air, water, and fire) and questions how alchemists determine what substances are elements.

1665 .... British scientist Robert Hooke publishes *Micrographia,* in which he describes pioneering applications of the optical microscope in chemistry, botany, and other scientific fields.

1667 .... The work of German alchemist Johann Joachim Becher forms the basis of the phlogiston theory of heat.

1669 .... German alchemist Hennig Brand discovers the element phosphorous (P).

1678 .... Robert Hooke studies the action of springs and reports that the extension (or compression) of an elastic material takes place in direct proportion to the force exerted on the material.

1687 .... British physicist Sir Isaac Newton publishes *The Principia.* His work provides the mathematical foundations for understanding (from a classical physics perspective) the motion of almost everything in the physical universe.

1738 .... Swiss mathematician Daniel Bernoulli publishes *Hydrodynamica*. In this seminal work, he identifies the relationships between density, pressure, and velocity in flowing fluids.

1748 .... While conducting experiments with electricity, American statesman and scientist Benjamin Franklin coins the term *battery*.

1754 .... Scottish chemist Joseph Black discovers a new gaseous substance, which he calls "fixed air." Other scientists later identify it as carbon dioxide (CO_2).

1764 .... Scottish engineer James Watt greatly improves the Newcomen steam engine. Watt steam engines power the First Industrial Revolution.

1772 .... Scottish physician and chemist Daniel Rutherford isolates a new colorless gaseous substance, calling it "noxious air." Other scientists soon refer to the new gas as nitrogen (N_2).

1785 .... French scientist Charles-Augustin de Coulomb performs experiments that lead to the important law of electrostatics, later known as Coulomb's law.

1789 .... French chemist Antoine-Laurent Lavoisier publishes *Treatise of Elementary Chemistry*, the first modern textbook on chemistry. Lavoisier also promotes the caloric theory of heat.

1800 .... Italian physicist Count Alessandro Volta invents the voltaic pile. His device is the forerunner of the modern electric battery.

1803 .... British schoolteacher and chemist John Dalton revives the atomic theory of matter. From his experiments, he concludes that all matter consists of combinations of atoms and that all the atoms of a particular element are identical.

1807 .... British chemist Sir Humphry Davy discovers the element potassium (K) while experimenting with caustic potash (KOH). Potassium is the first metal isolated by the process of electrolysis.

1811 .... Italian physicist Amedeo Avogadro proposes that equal volumes of different gases under the same conditions of pressure and temperature contain the same number of molecules. Scientists call this important hypothesis Avogadro's law.

1820 .... Danish physicist Hans Christian Ørsted discovers a relationship between magnetism and electricity.

1824 .... French military engineer Sadi Carnot publishes *Reflections on the Motive Power of Fire.* Despite the use of caloric theory, his work correctly identifies the general thermodynamic principles that govern the operation and efficiency of all heat engines.

1826 .... French scientist André-Marie Ampère experimentally formulates the relationship between electricity and magnetism.

1827 .... Experiments performed by German physicist George Simon Ohm indicate a fundamental relationship among voltage, current, and resistance.

1828 .... Swedish chemist Jöns Jacob Berzelius discovers the element thorium (Th).

1831 .... British experimental scientist Michael Faraday discovers the principle of electromagnetic induction. This principle is the basis for the electric dynamo.

Independent of Faraday, the American physicist Joseph Henry publishes a paper describing the electric motor (essentially a reverse dynamo).

1841 .... German physicist and physician Julius Robert von Mayer states the conservation of energy principle, namely that energy can neither be created nor destroyed.

1847 .... British physicist James Prescott Joule experimentally determines the mechanical equivalent of heat. Joule's work is a major step in developing the modern science of thermodynamics.

1866 .... Swedish scientist-industrialist Alfred Nobel finds a way to stabilize nitroglycerin and calls the new chemical explosive mixture dynamite.

1869 .... Russian chemist Dmitri Mendeleev introduces a periodic listing of the 63 known chemical elements in *Principles of Chemistry.* His periodic table includes gaps for elements predicted but not yet discovered.

American printer John W. Hyatt formulates celluloid, a flammable thermoplastic material made from a mixture of cellulose nitrate, alcohol, and camphor.

1873 .... Scottish mathematician and theoretical physicist James Clerk Maxwell publishes *Treatise on Electricity and Magnetism.*

1876 .... American physicist and chemist Josiah Willard Gibbs publishes *On the Equilibrium of Heterogeneous Substances.* This compendium forms the theoretical foundation of physical chemistry.

1884 .... Swedish chemist Svante Arrhenius proposes that electrolytes split or dissociate into electrically opposite positive and negative ions.

1888 .... German physicist Heinrich Rudolf Hertz produces and detects radio waves.

1895 .... German physicist Wilhelm Conrad Roentgen discovers X-rays.

1896 .... While investigating the properties of uranium salt, French physicist Antoine-Henri Becquerel discovers radioactivity.

1897 .... British physicist Sir Joseph John Thomson performs experiments that demonstrate the existence of the electron—the first subatomic particle discovered.

1898 .... French scientists Pierre and (Polish-born) Marie Curie announce the discovery of two new radioactive elements, polonium (Po) and radium (Ra).

1900 .... German physicist Max Planck postulates that blackbodies radiate energy only in discrete packets (or quanta) rather than continuously. His hypothesis marks the birth of quantum theory.

1903 .... New Zealand–born British physicist Baron (Ernest) Rutherford and British radiochemist Frederick Soddy propose the law of radioactive decay.

1904 .... German physicist Ludwig Prandtl revolutionizes fluid mechanics by introducing the concept of the boundary layer and its role in fluid flow.

1905 .... Swiss-German-American physicist Albert Einstein publishes the special theory of relativity, including the famous mass-energy equivalence formula ($E = mc^2$).

1907 .... Belgian-American chemist Leo Baekeland formulates bakelite. This synthetic thermoplastic material ushers in the age of plastics.

1911.................... Ernest Rutherford proposes the concept of the atomic nucleus based on the startling results of an alpha particle–gold foil scattering experiment.

1912 .... German physicist Max von Laue discovers that X-rays are diffracted by crystals.

1913 .... Danish physicist Niels Bohr presents his theoretical model of the hydrogen atom—a brilliant combination of atomic theory with quantum physics.

Frederick Soddy proposes the existence of isotopes.

1914 .... British physicist Henry Moseley measures the characteristic X-ray lines of many chemical elements.

1915 .... Albert Einstein presents his general theory of relativity, which relates gravity to the curvature of space-time.

1919 .... Ernest Rutherford bombards nitrogen (N) nuclei with alpha particles, causing the nitrogen nuclei to transform into oxygen (O) nuclei and to emit protons (hydrogen nuclei).

British physicist Francis Aston uses the newly invented mass spectrograph to identify more than 200 naturally occurring isotopes.

1923 .... American physicist Arthur Holly Compton conducts experiments involving X-ray scattering that demonstrate the particle nature of energetic photons.

1924 .... French physicist Louis-Victor de Broglie proposes the particle-wave duality of matter.

1926 .... Austrian physicist Erwin Schrödinger develops quantum wave mechanics to describe the dual wave-particle nature of matter.

1927 .... German physicist Werner Heisenberg introduces his uncertainty principle.

1929 .... American astronomer Edwin Hubble announces that his observations of distant galaxies suggest an expanding universe.

1932 .... British physicist Sir James Chadwick discovers the neutron.

British physicist Sir John Cockcroft and Irish physicist Ernest Walton use a linear accelerator to bombard lithium (Li) with energetic protons, producing the first artificial disintegration of an atomic nucleus.

American physicist Carl D. Anderson discovers the positron.

1934 .... Italian-American physicist Enrico Fermi proposes a theory of beta decay that includes the neutrino. He also starts to bombard uranium with neutrons and discovers the phenomenon of slow neutrons.

1938 .... German chemists Otto Hahn and Fritz Strassmann bombard uranium with neutrons and detect the presence of lighter elements. Austrian physicist Lise Meitner and Austrian-British physicist Otto Frisch review Hahn's work and conclude in early 1939 that the German chemists had split the atomic nucleus, achieving neutron-induced nuclear fission.

E.I. du Pont de Nemours & Company introduces a new thermoplastic material called nylon.

1941.................. American nuclear scientist Glenn T. Seaborg and his associates use the cyclotron at the University of California, Berkeley, to synthesize plutonium (Pu).

1942 .... Modern nuclear age begins when Enrico Fermi's scientific team at the University of Chicago achieves the first self-sustained, neutron-induced fission chain reaction at Chicago Pile One (CP-1), a uranium-fueled, graphite-moderated atomic pile (reactor).

1945 .... American scientists successfully detonate the world's first nuclear explosion, a plutonium-implosion device code-named Trinity.

1947 .... American physicists John Bardeen, Walter Brattain, and William Shockley invent the transistor.

1952 .... A consortium of 11 founding countries establishes CERN, the European Organization for Nuclear Research, at a site near Geneva, Switzerland.

United States tests the world's first thermonuclear device (hydrogen bomb) at the Enewetak Atoll in the Pacific Ocean. Code-named Ivy Mike, the experimental device produces a yield of 10.4 megatons.

1964 .... German-American physicist Arno Allen Penzias and American physicist Robert Woodrow Wilson detect the cosmic microwave background (CMB).

1967 .... German-American physicist Hans Albrecht Bethe receives the 1967 Nobel Prize in physics for his theory of thermonuclear reactions being responsible for energy generation in stars.

1969 .... On July 20, American astronauts Neil Armstrong and Edwin "Buzz" Aldrin successfully land on the Moon as part of NASA's *Apollo 11* mission.

1972 .... NASA launches the *Pioneer 10* spacecraft. It eventually becomes the first human-made object to leave the solar system on an interstellar trajectory

1985 .... American chemists Robert F. Curl, Jr., and Richard E. Smalley, collaborating with British astronomer Sir Harold W. Kroto, discover the buckyball, an allotrope of pure carbon.

1996 .... Scientists at CERN (near Geneva, Switzerland) announce the creation of antihydrogen, the first human-made antimatter atom.

1998 .... Astrophysicists investigating very distant Type 1A supernovae discover that the universe is expanding at an accelerated rate. Scientists coin the term *dark energy* in their efforts to explain what these observations physically imply.

2001 .... American physicist Eric A. Cornell, German physicist Wolfgang Ketterle, and American physicist Carl E. Wieman share the 2001 Nobel Prize in physics for their fundamental studies of the properties of Bose-Einstein condensates.

2005 .... Scientists at the Lawrence Livermore National Laboratory (LLNL) in California and the Joint Institute for Nuclear Research (JINR) in Dubna, Russia, perform collaborative experiments that establish the existence of super-heavy element 118, provisionally called ununoctium (Uuo).

2008 .... An international team of scientists inaugurates the world's most powerful particle accelerator, the Large Hadron Collider (LHC), located at the CERN laboratory near Geneva, Switzerland.

2009 .... British scientist Charles Kao, American scientist Willard Boyle, and American scientist George Smith share the 2009 Nobel Prize in physics for their pioneering efforts in fiber optics and imaging semiconductor devices, developments that unleashed the information technology revolution.

2010 .... Element 112 is officially named Copernicum (Cn) by the IUPAC in honor of Polish astronomer Nicholas Copernicus (1473–1543), who championed heliocentric cosmology.

Scientists at the Joint Institute for Nuclear Research in Dubna, Russia, announce the synthesis of element 117 (ununseptium [Uus]) in early April.

Glossary

absolute zero the lowest possible temperature; equal to 0 kelvin (K) (–459.67°F, –273.15°C)

acceleration (a) rate at which the velocity of an object changes with time

accelerator device for increasing the velocity and energy of charged elementary particles

acid substance that produces hydrogen ions (H^+) when dissolved in water

actinoid (formerly actinide) series of heavy metallic elements beginning with element 89 (actinium) and continuing through element 103 (lawrencium)

activity measure of the rate at which a material emits nuclear radiations

air overall mixture of gases that make up Earth's atmosphere

alchemy mystical blend of sorcery, religion, and prescientific chemistry practiced in many early societies around the world

alloy solid solution (compound) or homogeneous mixture of two or more elements, at least one of which is an elemental metal

alpha particle (α) positively charged nuclear particle emitted from the nucleus of certain radioisotopes when they undergo decay; consists of two protons and two neutrons bound together

alternating current (AC) electric current that changes direction periodically in a circuit

American customary system of units (also American system) used primarily in the United States; based on the foot (ft), pound-mass (lbm), pound-force (lbf), and second (s). Peculiar to this system is the artificial construct (based on Newton's second law) that one pound-force equals one pound-mass (lbm) at sea level on Earth

ampere (A) SI unit of electric current

anode positive electrode in a battery, fuel cell, or electrolytic cell; oxidation occurs at anode

antimatter matter in which the ordinary nuclear particles are replaced by corresponding antiparticles

Archimedes principle the fluid mechanics rule that states that the buoyant (upward) force exerted on a solid object immersed in a fluid equals the weight of the fluid displaced by the object

atom smallest part of an element, indivisible by chemical means; consists of a dense inner core (nucleus) that contains protons and neutrons and a cloud of orbiting electrons

atomic mass *See* **relative atomic mass**

atomic mass unit (amu) 1/12 mass of carbon's most abundant isotope, namely carbon-12

atomic number (Z) total number of protons in the nucleus of an atom and its positive charge

atomic weight the mass of an atom relative to other atoms. *See also* **relative atomic mass**

battery electrochemical energy storage device that serves as a source of direct current or voltage

becquerel (Bq) SI unit of radioactivity; one disintegration (or spontaneous nuclear transformation) per second. *Compare with* **curie**

beta particle (β) elementary particle emitted from the nucleus during radioactive decay; a negatively charged beta particle is identical to an electron

big bang theory in cosmology concerning the origin of the universe; postulates that about 13.7 billion years ago, an initial singularity experienced a very large explosion that started space and time. Astrophysical observations support this theory and suggest that the universe has been expanding at different rates under the influence of gravity, dark matter, and dark energy

blackbody perfect emitter and perfect absorber of electromagnetic radiation; radiant energy emitted by a blackbody is a function only of the emitting object's absolute temperature

black hole incredibly compact, gravitationally collapsed mass from which nothing can escape

boiling point temperature (at a specified pressure) at which a liquid experiences a change of state into a gas

Bose-Einstein condensate (BEC) state of matter in which extremely cold atoms attain the same quantum state and behave essentially as a large "super atom"

boson general name given to any particle with a spin of an integral number (0, 1, 2, etc.) of quantum units of angular momentum. Carrier particles of all interactions are bosons. *See also* **carrier particle**

brass alloy of copper (Cu) and zinc (Zn)

British thermal unit (Btu) amount of heat needed to raise the temperature of 1 lbm of water 1°F at normal atmospheric pressure; 1 Btu = 1,055 J = 252 cal

bronze alloy of copper (Cu) and tin (Sn)

calorie (cal) quantity of heat; defined as the amount needed to raise one gram of water 1°C at normal atmospheric pressure; 1 cal = 4.1868 J = 0.004 Btu

carbon dioxide (CO_2) colorless, odorless, noncombustible gas present in Earth's atmosphere

Carnot cycle ideal reversible thermodynamic cycle for a theoretical heat engine; represents the best possible thermal efficiency of any heat engine operating between two absolute temperatures (T_1 and T_2)

carrier particle within the standard model, gluons are carrier particles for strong interactions; photons are carrier particles of electromagnetic interactions; and the W and Z bosons are carrier particles for weak interactions. *See also* **standard model**

catalyst substance that changes the rate of a chemical reaction without being consumed or changed by the reaction

cathode negative electrode in a battery, fuel cell, electrolytic cell, or electron (discharge) tube through which a primary stream of electrons enters a system

chain reaction reaction that stimulates its own repetition. *See also* **nuclear chain reaction**

change of state the change of a substance from one physical state to another; the atoms or molecules are structurally rearranged without experiencing a change in composition. Sometimes called change of phase or phase transition

charged particle elementary particle that carries a positive or negative electric charge

chemical bond(s) force(s) that holds atoms together to form stable configurations of molecules

chemical property characteristic of a substance that describes the manner in which the substance will undergo a reaction with another substance, resulting in a change in chemical composition. *Compare with* **physical property**

chemical reaction involves changes in the electron structure surrounding the nucleus of an atom; a dissociation, recombination, or rearrangement of atoms. During a chemical reaction, one or more kinds of matter (called reactants) are transformed into one or several new kinds of matter (called products)

color charge in the standard model, the charge associated with strong interactions. Quarks and gluons have color charge and thus participate in strong interactions. Leptons, photons, W bosons, and Z bosons do not have color charge and consequently do not participate in strong interactions. *See also* **standard model**

combustion chemical reaction (burning or rapid oxidation) between a fuel and oxygen that generates heat and usually light

composite materials human-made materials that combine desirable properties of several materials to achieve an improved substance; includes combinations of metals, ceramics, and plastics with built-in strengthening agents

compound pure substance made up of two or more elements chemically combined in fixed proportions

compressible flow fluid flow in which density changes cannot be neglected

compression condition when an applied external force squeezes the atoms of a material closer together. *Compare with* **tension**

concentration for a solution, the quantity of dissolved substance per unit quantity of solvent

condensation change of state process by which a vapor (gas) becomes a liquid. *The opposite of* **evaporation**

conduction (thermal) transport of heat through an object by means of a temperature difference from a region of higher temperature to a region of lower temperature. *Compare with* **convection**

conservation of mass and energy Einstein's special relativity principle stating that energy (E) and mass (m) can neither be created nor destroyed, but are interchangeable in accordance with the equation $E = mc^2$, where c represents the speed of light

convection fundamental form of heat transfer characterized by mass motions within a fluid resulting in the transport and mixing of the properties of that fluid

coulomb (C) SI unit of electric charge; equivalent to quantity of electric charge transported in one second by a current of one ampere

covalent bond the chemical bond created within a molecule when two or more atoms share an electron

creep slow, continuous, permanent deformation of solid material caused by a constant tensile or compressive load that is less than the load necessary for the material to give way (yield) under pressure. *See also* **plastic deformation**

crystal a solid whose atoms are arranged in an orderly manner, forming a distinct, repetitive pattern

curie (Ci) traditional unit of radioactivity equal to 37 billion (37×10^9) disintegrations per second. *Compare with* **becquerel**

current (I) flow of electric charge through a conductor

dark energy a mysterious natural phenomenon or unknown cosmic force thought responsible for the observed acceleration in the rate of expansion of the universe. Astronomical observations suggest dark energy makes up about 72 percent of the universe

dark matter (nonbaryonic matter) exotic form of matter that emits very little or no electromagnetic radiation. It experiences no measurable interaction with ordinary (baryonic) matter but somehow accounts for the observed structure of the universe. It makes up about 23 percent of the content of the universe, while ordinary matter makes up less than 5 percent

density (ρ) mass of a substance per unit volume at a specified temperature

deposition direct transition of a material from the gaseous (vapor) state to the solid state without passing through the liquid phase. *Compare with* **sublimation**

dipole magnet any magnet with one north and one south pole

direct current (DC) electric current that always flows in the same direction through a circuit

elastic deformation temporary change in size or shape of a solid due to an applied force (stress); when force is removed the solid returns to its original size and shape

elasticity ability of a body that has been deformed by an applied force to return to its original shape when the force is removed

elastic modulus a measure of the stiffness of a solid material; defined as the ratio of stress to strain

electricity flow of energy due to the motion of electric charges; any physical effect that results from the existence of moving or stationary electric charges

electrode conductor (terminal) at which electricity passes from one medium into another; positive electrode is the *anode;* negative electrode is the *cathode*

electrolyte a chemical compound that, in an aqueous (water) solution, conducts an electric current

electromagnetic radiation (EMR) oscillating electric and magnetic fields that propagate at the speed of light. Includes in order of increasing frequency and energy: radio waves, radar waves, infrared (IR) radiation, visible light, ultraviolet radiation, X-rays, and gamma rays

electron (e) stable elementary particle with a unit negative electric charge (1.602×10^{-19} C). Electrons form an orbiting cloud, or shell, around the positively charged atomic nucleus and determine an atom's chemical properties

electron volt (eV) energy gained by an electron as it passes through a potential difference of one volt; one electron volt has an energy equivalence of 1.519×10^{-22} Btu = 1.602×10^{-19} J

element pure chemical substance indivisible into simpler substances by chemical means; all the atoms of an element have the same number of protons in the nucleus and the same number of orbiting electrons, although the number of neutrons in the nucleus may vary

elementary particle a fundamental constituent of matter; the basic atomic model suggests three elementary particles: the proton, neutron, and electron. *See also* **fundamental particle**

endothermic reaction chemical reaction requiring an input of energy to take place. *Compare with* **exothermic reaction**

energy (E) capacity to do work; appears in many different forms, such as mechanical, thermal, electrical, chemical, and nuclear

entropy (S) measure of disorder within a system; as entropy increases, energy becomes less available to perform useful work

evaporation physical process by which a liquid is transformed into a gas (vapor) at a temperature below the boiling point of the liquid. *Compare with* **sublimation**

excited state state of a molecule, atom, electron, or nucleus when it possesses more than its normal energy. *Compare with* **ground state**

exothermic reaction chemical reaction that releases energy as it takes place. *Compare with* **endothermic reaction**

fatigue weakening or deterioration of metal or other material that occurs under load, especially under repeated cyclic or continued loading

fermion general name scientists give to a particle that is a matter constituent. Fermions are characterized by spin in odd half-integer quantum units (namely, 1/2, 3/2, 5/2, etc.); quarks, leptons, and baryons are all fermions

fission (nuclear) splitting of the nucleus of a heavy atom into two lighter nuclei accompanied by the release of a large amount of energy as well as neutrons, X-rays, and gamma rays

flavor in the standard model, quantum number that distinguishes different types of quarks and leptons. *See also* **quark; lepton**

fluid mechanics scientific discipline that deals with the behavior of fluids (both gases and liquids) at rest (fluid statics) and in motion (fluid dynamics)

foot-pound (force) ($ft\text{-}lb_{force}$) unit of work in American customary system of units; 1 $ft\text{-}lb_{force}$ = 1.3558 J

force (F) the cause of the acceleration of material objects as measured by the rate of change of momentum produced on a free body. Force is a vector quantity mathematically expressed by Newton's second law of motion: force = mass × acceleration

freezing point the temperature at which a substance experiences a change from the liquid state to the solid state at a specified pressure; at this temperature, the solid and liquid states of a substance can coexist in equilibrium. *Synonymous with* **melting point**

fundamental particle particle with no internal substructure; in the standard model, any of the six types of quarks or six types of leptons and their antiparticles. Scientists postulate that all other particles are made from a combination of quarks and leptons. *See also* **elementary particle**

fusion (nuclear) nuclear reaction in which lighter atomic nuclei join together (fuse) to form a heavier nucleus, liberating a great deal of energy

g acceleration due to gravity at sea level on Earth; approximately 32.2 ft/s^2 (9.8 m/s^2)

gamma ray (γ) high-energy, very short–wavelength photon of electromagnetic radiation emitted by a nucleus during certain nuclear reactions or radioactive decay

gas state of matter characterized as an easily compressible fluid that has neither a constant volume nor a fixed shape; a gas assumes the total size and shape of its container

gravitational lensing bending of light from a distant celestial object by a massive (gravitationally influential) foreground object

ground state state of a nucleus, atom, or molecule at its lowest (normal) energy level

hadron any particle (such as a baryon) that exists within the nucleus of an atom; made up of quarks and gluons, hadrons interact with the strong force

half-life (radiological) time in which half the atoms of a particular radioactive isotope disintegrate to another nuclear form

heat energy transferred by a temperature difference or thermal process. *Compare* **work**

heat capacity (c) amount of heat needed to raise the temperature of an object by one degree

heat engine thermodynamic system that receives energy in the form of heat and that, in the performance of energy transformation on a working fluid, does work. Heat engines function in thermodynamic cycles

hertz (Hz) SI unit of frequency; equal to one cycle per second

high explosive (HE) energetic material that detonates (rather than burns); the rate of advance of the reaction zone into the unreacted material exceeds the velocity of sound in the unreacted material

horsepower (hp) American customary system unit of power; 1 hp = 550 $\text{ft-lb}_{\text{force}}/\text{s}$ = 746 W

hydraulic operated, moved, or affected by liquid used to transmit energy

hydrocarbon organic compound composed of only carbon and hydrogen atoms

ideal fluid *See* **perfect fluid**

ideal gas law important physical principle: $P\ V = n\ R_u\ T$, where P is pressure, V is volume, T is temperature, n is the number of moles of gas, and R_u is the universal gas constant

incompressible flow fluid flow in which density changes can be neglected. *Compare with* **compressible flow**

inertia resistance of a body to a change in its state of motion

infrared (IR) radiation that portion of the electromagnetic (EM) spectrum lying between the optical (visible) and radio wavelengths

International System of units *See* **SI unit system**

inviscid fluid perfect fluid that has zero coefficient of viscosity. *See* **perfect fluid**

ion atom or molecule that has lost or gained one or more electrons, so that the total number of electrons does not equal the number of protons

ionic bond formed when one atom gives up at least one outer electron to another atom, creating a chemical bond–producing electrical attraction between the atoms

isotope atoms of the same chemical element but with different numbers of neutrons in their nucleus

joule (J) basic unit of energy or work in the SI unit system; 1 J = 0.2388 calorie = 0.00095 Btu

kelvin (K) SI unit of absolute thermodynamic temperature

kinetic energy (KE) energy due to motion

lepton fundamental particle of matter that does not participate in strong interactions; in the standard model, the three charged leptons are the electron (e), the muon (μ), and the tau (τ) particle; the three neutral leptons are the electron neutrino (ν_e), the muon neutrino (ν_μ), and the tau neutrino (ν_τ). A corresponding set of antiparticles also exists. *See also* **standard model**

light-year (ly) distance light travels in one year; 1 ly ≈ 5.88×10^{12} miles (9.46×10^{12} km)

liquid state of matter characterized as a relatively incompressible flowing fluid that maintains an essentially constant volume but assumes the shape of its container

liter (l or L) SI unit of volume; 1 L = 0.264 gal

magnet material or device that exhibits magnetic properties capable of causing the attraction or repulsion of another magnet or the attraction of certain ferromagnetic materials such as iron

manufacturing process of transforming raw material(s) into a finished product, especially in large quantities

mass (m) property that describes how much material makes up an object and gives rise to an object's inertia

mass number *See* **relative atomic mass**

mass spectrometer instrument that measures relative atomic masses and relative abundances of isotopes

material tangible substance (chemical, biological, or mixed) that goes into the makeup of a physical object

mechanics branch of physics that deals with the motions of objects

melting point temperature at which a substance experiences a change from the solid state to the liquid state at a specified pressure; at this temperature, the solid and liquid states of a substance can coexist in equilibrium. *Synonymous with* **freezing point**

metallic bond chemical bond created as many atoms of a metallic substance share the same electrons

meter (m) fundamental SI unit of length; 1 meter = 3.281 feet. British spelling *metre*

metric system *See* **SI unit system**

metrology science of dimensional measurement; sometimes includes the science of weighing

microwave (radiation) comparatively short-wavelength electromagnetic (EM) wave in the radio frequency portion of the EM spectrum

mirror matter *See* **antimatter**

mixture a combination of two or more substances, each of which retains its own chemical identity

molarity (M) concentration of a solution expressed as moles of solute per kilogram of solvent

mole (mol) SI unit of the amount of a substance; defined as the amount of substance that contains as many elementary units as there are atoms in 0.012 kilograms of carbon-12, a quantity known as Avogadro's number (N_A), which has a value of about 6.022×10^{23} molecules/mole

molecule smallest amount of a substance that retains the chemical properties of the substance; held together by chemical bonds, a molecule can consist of identical atoms or different types of atoms

monomer substance of relatively low molecular mass; any of the small molecules that are linked together by covalent bonds to form a polymer

natural material material found in nature, such as wood, stone, gases, and clay

neutrino (ν) lepton with no electric charge and extremely low (if not zero) mass; three known types of neutrinos are the electron neutrino (ν_e), the muon neutrino (ν_μ), and the tau neutrino (ν_τ). *See also* **lepton**

neutron (n) an uncharged elementary particle found in the nucleus of all atoms except ordinary hydrogen. Within the standard model, the neutron is a baryon with zero electric charge consisting of two down (d) quarks and one up (u) quark. *See also* **standard model**

newton (N) The SI unit of force; 1 N = 0.2248 lbf

nuclear chain reaction occurs when a fissionable nuclide (such as plutonium-239) absorbs a neutron, splits (or fissions), and releases several neutrons along with energy. A fission chain reaction is self-sustaining when (on average) at least one released neutron per fission event survives to create another fission reaction

nuclear energy energy released by a nuclear reaction (fission or fusion) or by radioactive decay

nuclear radiation particle and electromagnetic radiation emitted from atomic nuclei as a result of various nuclear processes, such as radioactive decay and fission

nuclear reaction reaction involving a change in an atomic nucleus, such as fission, fusion, neutron capture, or radioactive decay

nuclear reactor device in which a fission chain reaction can be initiated, maintained, and controlled

nuclear weapon precisely engineered device that releases nuclear energy in an explosive manner as a result of nuclear reactions involving fission, fusion, or both

nucleon constituent of an atomic nucleus; a proton or a neutron

nucleus (plural: nuclei) small, positively charged central region of an atom that contains essentially all of its mass. All nuclei contain both protons and neutrons except the nucleus of ordinary hydrogen, which consists of a single proton

nuclide general term applicable to all atomic (isotopic) forms of all the elements; nuclides are distinguished by their atomic number, relative mass number (atomic mass), and energy state

ohm (Ω) SI unit of electrical resistance

oxidation chemical reaction in which oxygen combines with another substance, and the substance experiences one of three processes: (1) the gaining of oxygen, (2) the loss of hydrogen, or (3) the loss of electrons. In these reactions, the substance being "oxidized" loses electrons and forms positive ions. *Compare with* **reduction**

oxidation-reduction (redox) reaction chemical reaction in which electrons are transferred between species or in which atoms change oxidation number

particle minute constituent of matter, generally one with a measurable mass

pascal (Pa) SI unit of pressure; 1 Pa = 1 N/m^2 = 0.000145 psi

Pascal's principle when an enclosed (static) fluid experiences an increase in pressure, the increase is transmitted throughout the fluid; the physical principle behind all hydraulic systems

Pauli exclusion principle postulate that no two electrons in an atom can occupy the same quantum state at the same time; also applies to protons and neutrons

perfect fluid hypothesized fluid primarily characterized by a lack of viscosity and usually by incompressibility

perfect gas law *See* **ideal gas law**

periodic table list of all the known elements, arranged in rows (periods) in order of increasing atomic numbers and columns (groups) by similar physical and chemical characteristics

phase one of several different homogeneous materials present in a portion of matter under study; the set of states of a large-scale (macroscopic) physical system having relatively uniform physical properties and chemical composition

phase transition *See* **change of state**

photon A unit (or particle) of electromagnetic radiation that carries a quantum (packet) of energy that is characteristic of the particular radiation. Photons travel at the speed of light and have an effective momentum, but no mass or electrical charge. In the standard model, a photon is the carrier particle of electromagnetic radiation

photovoltaic cell *See* **solar cell**

physical property characteristic quality of a substance that can be measured or demonstrated without changing the composition or chemical identity of the substance, such as temperature and density. *Compare with* **chemical property**

Planck's constant (h) fundamental physical constant describing the extent to which quantum mechanical behavior influences nature. Equals the ratio of a photon's energy (E) to its frequency (ν), namely: $h = E/\nu = 6.626 \times 10^{-34}$ J-s (6.282×10^{-37} Btu-s). *See also* **uncertainty principle**

plasma electrically neutral gaseous mixture of positive and negative ions; called the fourth state of matter

plastic deformation permanent change in size or shape of a solid due to an applied force (stress)

plasticity tendency of a loaded body to assume a (deformed) state other than its original state when the load is removed

plastics synthesized family of organic (mainly hydrocarbon) polymer materials used in nearly every aspect of modern life

pneumatic operated, moved, or effected by a pressurized gas (typically air) that is used to transmit energy

polymer very large molecule consisting of a number of smaller molecules linked together repeatedly by covalent bonds, thereby forming long chains

positron (e^+ or β^+) elementary antimatter particle with the mass of an electron but charged positively

pound-force (lbf) basic unit of force in the American customary system; 1 lbf = 4.448 N

pound-mass (lbm) basic unit of mass in the American customary system; 1 lbm = 0.4536 kg

power rate with respect to time at which work is done or energy is transformed or transferred to another location; 1 hp = 550 ft-lb_{force}/s = 746 W

pressure (P) the normal component of force per unit area exerted by a fluid on a boundary; 1 psi = 6,895 Pa

product substance produced by or resulting from a chemical reaction

proton (p) stable elementary particle with a single positive charge. In the the standard model, the proton is a baryon with an electric charge of +1; it consists of two up (u) quarks and one down (d) quark. *See also* **standard model**

quantum mechanics branch of physics that deals with matter and energy on a very small scale; physical quantities are restricted to discrete values and energy to discrete packets called quanta

quark fundamental matter particle that experiences strong-force interactions. The six flavors of quarks in order of increasing mass are up (u), down (d), strange (s), charm (c), bottom (b), and top (t)

radiation heat transfer The transfer of heat by electromagnetic radiation that arises due to the temperature of a body; can takes place in and through a vacuum

radioactive isotope unstable isotope of an element that decays or disintegrates spontaneously, emitting nuclear radiation; also called radioisotope

radioactivity spontaneous decay of an unstable atomic nucleus, usually accompanied by the emission of nuclear radiation, such as alpha particles, beta particles, gamma rays, or neutrons

radio frequency (RF) a frequency at which electromagnetic radiation is useful for communication purposes; specifically, a frequency above 10,000 hertz (Hz) and below 3×10^{11}Hz

rankine (R) American customary unit of absolute temperature. *See also* **kelvin (K)**

reactant original substance or initial material in a chemical reaction

reduction portion of an oxidation-reduction (redox) reaction in which there is a gain of electrons, a gain in hydrogen, or a loss of oxygen. *See also* **oxidation-reduction (redox) reaction**

relative atomic mass (A) total number of protons and neutrons (nucleons) in the nucleus of an atom. Previously called *atomic mass* or *atomic mass number. See also* **atomic mass unit**

residual electromagnetic effect force between electrically neutral atoms that leads to the formation of molecules

residual strong interaction interaction responsible for the nuclear binding force—that is, the strong force holding hadrons (protons and neutrons) together in the atomic nucleus. *See also* **strong force**

resilience property of a material that enables it to return to its original shape and size after deformation

resistance (R) the ratio of the voltage (V) across a conductor to the electric current (I) flowing through it

scientific notation A method of expressing powers of 10 that greatly simplifies writing large numbers; for example, $3 \times 10^6 = 3{,}000{,}000$

SI unit system international system of units (the metric system), based upon the meter (m), kilogram (kg), and second (s) as the fundamental units of length, mass, and time, respectively

solar cell (photovoltaic cell) a semiconductor direct energy conversion device that transforms sunlight into electric energy

solid state of matter characterized by a three-dimensional regularity of structure; a solid is relatively incompressible, maintains a fixed volume, and has a definitive shape

solution When scientists dissolve a substance in a pure liquid, they refer to the dissolved substance as the *solute* and the host pure liquid as the *solvent*. They call the resulting intimate mixture the solution

spectroscopy study of spectral lines from various atoms and molecules; emission spectroscopy infers the material composition of the objects that emitted the light; absorption spectroscopy infers the composition of the intervening medium

speed of light *(c)* speed at which electromagnetic radiation moves through a vacuum; regarded as a universal constant equal to 186,283.397 mi/s (299,792.458 km/s)

stable isotope isotope that does not undergo radioactive decay

standard model contemporary theory of matter, consisting of 12 fundamental particles (six quarks and six leptons), their respective antiparticles, and four force carriers (gluons, photons, W bosons, and Z bosons)

state of matter form of matter having physical properties that are quantitatively and qualitatively different from other states of matter; the three more common states on Earth are solid, liquid, and gas

steady state condition of a physical system in which parameters of importance (fluid velocity, temperature, pressure, etc.) do not vary significantly with time

strain the change in the shape or dimensions (volume) of an object due to applied forces; longitudinal, volume, and shear are the three basic types of strain

stress applied force per unit area that causes an object to deform (experience strain); the three basic types of stress are compressive (or tensile) stress, hydrostatic pressure, and shear stress

string theory theory of quantum gravity that incorporates Einstein's general relativity with quantum mechanics in an effort to explain space-time phenomena on the smallest imaginable scales; vibrations of incredibly tiny stringlike structures form quarks and leptons

strong force In the standard model, the fundamental force between quarks and gluons that makes them combine to form hadrons, such as protons and neutrons; also holds hadrons together in a nucleus. *See also* **standard model**

subatomic particle any particle that is small compared to the size of an atom

sublimation direct transition of a material from the solid state to the gaseous (vapor) state without passing through the liquid phase. *Compare with* **deposition**

superconductivity the ability of a material to conduct electricity without resistance at a temperature above absolute zero

temperature (T) thermodynamic property that serves as a macroscopic measure of atomic and molecular motions within a substance; heat naturally flows from regions of higher temperature to regions of lower temperature

tension condition when applied external forces pull atoms of a material farther apart. *Compare with* **compression**

thermal conductivity (k) intrinsic physical property of a substance; a material's ability to conduct heat as a consequence of molecular motion

thermodynamics branch of science that treats the relationships between heat and energy, especially mechanical energy

thermodynamic system collection of matter and space with boundaries defined in such a way that energy transfer (as work and heat) from and to the system across these boundaries can be easily identified and analyzed

thermometer instrument or device for measuring temperature

toughness ability of a material (especially a metal) to absorb energy and deform plastically before fracturing

transmutation transformation of one chemical element into a different chemical element by a nuclear reaction or series of reactions

transuranic element (isotope) human-made element (isotope) beyond uranium on the periodic table

ultraviolet (UV) radiation portion of the electromagnetic spectrum that lies between visible light and X-rays

uncertainty principle Heisenberg's postulate that places quantum-level limits on how accurately a particle's momentum *(p)* and position *(x)* can be simultaneously measured. Planck's constant (h) expresses this uncertainty as $\Delta x \times \Delta p \geq h/2\pi$

U.S. customary system of units *See* **American customary system of units**

vacuum relative term used to indicate the absence of gas or a region in which there is a very low gas pressure

valence electron electron in the outermost shell of an atom

van der Waals force generally weak interatomic or intermolecular force caused by polarization of electrically neutral atoms or molecules

vapor gaseous state of a substance

velocity vector quantity describing the rate of change of position; expressed as length per unit of time

velocity of light (c) *See* **speed of light**

viscosity measure of the internal friction or flow resistance of a fluid when it is subjected to shear stress

volatile solid or liquid material that easily vaporizes; volatile material has a relatively high vapor pressure at normal temperatures

volt (V) SI unit of electric potential difference

volume (V) space occupied by a solid object or a mass of fluid (liquid or confined gas)

watt (W) SI unit of power (work per unit time); 1 W = 1 J/s = 0.00134 hp = 0.737 $\text{ft-lb}_{\text{force}}$/s

wavelength (λ) the mean distance between two adjacent maxima (or minima) of a wave

weak force fundamental force of nature responsible for various types of radioactive decay

weight (w) the force of gravity on a body; on Earth, product of the mass (m) of a body times the acceleration of gravity (g), namely $w = \text{m} \times g$

work (W) energy expended by a force acting though a distance. *Compare with* **heat**

X-ray penetrating form of electromagnetic (EM) radiation that occurs on the EM spectrum between ultraviolet radiation and gamma rays

Further Resources

BOOKS

Allcock, Harry R. *Introduction to Materials Chemistry.* New York: John Wiley & Sons, 2008. A college-level textbook that provides a basic treatment of the principles of chemistry upon which materials science depends.

Angelo, Joseph A., Jr. *Nuclear Technology.* Westport, Conn.: Greenwood Press, 2004. The book provides a detailed discussion of both military and civilian nuclear technology and includes impacts, issues, and future advances.

———. *Encyclopedia of Space and Astronomy.* New York: Facts On File, 2006. Provides a comprehensive treatment of major concepts in astronomy, astrophysics, planetary science, cosmology, and space technology.

Ball, Philip. *Designing the Molecular World: Chemistry at the Frontier.* Princeton, N.J.: Princeton University Press, 1996. Discusses many recent advances in modern chemistry, including nanotechnology and superconductor materials.

———. *Made to Measure: New Materials for the 21st Century.* Princeton, N.J.: Princeton University Press, 1998. Discusses how advanced new materials can significantly influence life in the 21st century.

Bensaude-Vincent, Bernadette, and Isabelle Stengers. *A History of Chemistry.* Cambridge, Mass.: Harvard University Press, 1996. Describes how chemistry emerged as a science and its impact on civilization.

Callister, William D., Jr. *Materials Science and Engineering: An Introduction.* 8th ed. New York: John Wiley & Sons, 2010. Intended primarily for engineers, technically knowledgeable readers will also benefit from this book's introductory treatment of metals, ceramics, polymers, and composite materials.

Charap, John M. *Explaining the Universe: The New Age of Physics.* Princeton, N.J.: Princeton University Press, 2004. Discusses the important discoveries in physics during the 20th century that are influencing civilization.

Close, Frank, et al. *The Particle Odyssey: A Journey to the Heart of the Matter.* New York: Oxford University Press, 2002. A well-illustrated and enjoyable tour of the subatomic world.

Cobb, Cathy, and Harold Goldwhite. *Creations of Fire: Chemistry's Lively History from Alchemy to the Atomic Age.* New York: Plenum Press, 1995. Uses historic circumstances and interesting individuals to describe the emergence of chemistry as a scientific discipline.

Feynman, Richard P. *QED: The Strange Theory of Light and Matter.* Princeton, N.J.: Princeton University Press, 2006. Written by an American Nobel laureate, addresses several key topics in modern physics.

Gordon, J. E. *The New Science of Strong Materials or Why You Don't Fall Through the Floor.* Princeton, N.J.: Princeton University Press, 2006. Discusses the science of structural materials in a manner suitable for both technical and lay audiences.

Hill, John W., and Doris K. Kolb. *Chemistry for Changing Times.* 11th ed. Upper Saddle River, N.J.: Pearson Prentice Hall, 2007. Readable college-level textbook that introduces all the basic areas of modern chemistry.

Krebs, Robert E. *The History and Use of Our Earth's Chemical Elements: A Reference Guide.* 2nd ed. Westport, Conn.: Greenwood Press, 2006. Provides a concise treatment of each of the chemical elements.

Levere, Trevor H. *Transforming Matter: A History of Chemistry from Alchemy to the Buckyball.* Baltimore: Johns Hopkins University Press, 2001. Provides an understandable overview of the chemical sciences from the early alchemists through modern times.

Lutgens, Frederick K., and Edward J. Tarbuck. *The Atmosphere: An Introduction to Meteorology.* 10th ed. Upper Saddle River, N.J.: Pearson Prentice Hall, 2007. Readable college-level textbook that discusses the atmosphere, meteorology, climate, and the physical properties of air.

Mackintosh, Ray, et al. *Nucleus: A Trip into the Heart of Matter.* Baltimore: Johns Hopkins University Press, 2001. Provides a technical though readable explanation of how modern scientists developed their current understanding of the atomic nucleus and the standard model.

Nicolaou, K. C., and Tamsyn Montagnon. *Molecules that Changed the World.* New York: John Wiley & Sons, 2008. Provides an interesting treatment of such important molecules as aspirin, camphor, glucose, quinine, and morphine.

Scerri, Eric R. *The Periodic Table: Its Story and Its Significance.* New York: Oxford University Press, 2007. Provides a detailed look at the periodic table and its iconic role in the practice of modern science.

Smith, William F., and Javad Hashemi. *Foundations of Materials Science and Engineering.* 5th ed. New York: McGraw-Hill, 2006. Provides scientists and engineers of all disciplines an introduction to materials science, including metals, ceramics, polymers, and composite materials. Technically knowledgeable laypersons will find the treatment of specific topics such as biological materials useful.

Strathern, Paul. *Mendeleyev's Dream: The Quest for the Elements.* New York: St. Martin's Press, 2001. Describes the intriguing history of chemistry from the early Greek philosophers to the 19th-century Russian chemist Dmitri Mendeleyev.

Thrower, Peter, and Thomas Mason. *Materials in Today's World.* 3rd ed. New York: McGraw-Hill Companies, 2007. Provides a readable introductory treatment of modern materials science, including biomaterials and nanomaterials.

Trefil, James, and Robert M. Hazen. *Physics Matters: An Introduction to Conceptual Physics.* New York: John Wiley & Sons, 2004. Highly-readable introductory college-level textbook that provides a good overview of physics from classical mechanics to relativity and cosmology. Laypersons will find the treatment of specific topics useful and comprehendible.

Zee, Anthony. *Quantum Field Theory in a Nutshell.* Princeton, N.J.: Princeton University Press, 2003. A reader-friendly treatment of the generally complex and profound physical concepts that constitute quantum field theory.

WEB SITES

To help enrich the content of this book and to make your investigation of matter more enjoyable, the following is a selective list of recommended Web sites. Many of the sites below will also lead to other interesting science-related locations on the Internet. Some sites provide unusual science learning opportunities (such as laboratory simulations) or in-depth educational resources.

American Chemical Society (ACS) is a congressionally chartered independent membership organization that represents professionals at all degree levels and in all fields of science involving chemistry. The ACS Web site includes educational resources for high school and college students. Available online. URL: http://portal.acs.org/portal/acs/corg/content. Accessed on February 12, 2010.

American Institute of Physics (AIP) is a not-for-profit corporation that promotes the advancement and diffusion of the knowledge of physics and its applications to human welfare. This Web site offers an enormous quantity of fascinating information about the history of physics from ancient Greece up to the present day. Available online. URL: http://www.aip.org/aip/. Accessed on February 12, 2010.

Chandra X-ray Observatory (CXO) is a space-based NASA astronomical observatory that observes the universe in the X-ray portion of the electromagnetic spectrum. This Web site contains contemporary information and educational materials about astronomy, astrophysics, and cosmology, including topics such as black holes, neutron stars, dark matter, and dark energy. Available online. URL: http://www.chandra.harvard.edu/. Accessed on February 12, 2010.

The ChemCollective is an online resource for learning about chemistry. Through simulations developed by the Department of Chemistry of Carnegie Mellon University (with funding from the National Science Foundation), a person gets the chance to safely mix chemicals without worrying about accidentally spilling them. Available online. URL: http://www.chemcollective.org/vlab/vlab.php. Accessed on February 12, 2010.

Chemical Heritage Foundation (CHF) maintains a rich and informative collection of materials that describe the history and heritage of the chemical and molecular sciences, technologies, and industries. Available online. URL: http://www.chemheritage.org/. Accessed on February 12, 2010.

Department of Defense (DOD) is responsible for maintaining armed forces of sufficient strength and technology to protect the United States and its citizens from all credible foreign threats. This Web site serves as an efficient access point to activities within the DOD, including those taking place within each of the individual armed services: the U.S. Army, U.S. Navy, U.S. Air Force, and U.S. Marines. As part of national security, the DOD sponsors a large amount of research and development, including activities in materials science, chemistry, physics, and nanotechnology. Available online. URL: http://www.defenselink.mil/. Accessed on February 12, 2010.

Department of Energy (DOE) is the single largest supporter of basic research in the physical sciences in the federal government of the United States. Topics found on this Web site include materials sciences, nanotechnology, energy sciences, chemical science, high-energy physics, and nuclear physics. The Web site also includes convenient links to all of the DOE's national laboratories. Available online. URL: http://energy.gov/. Accessed on February 12, 2010.

Fermi National Accelerator Laboratory (Fermilab) performs research that advances the understanding of the fundamental nature of matter and energy. Fermilab's Web site contains contemporary information about particle physics, the standard model, and the impact of particle physics on society. Available online. URL: http://www.fnal.gov/. Accessed on February 12, 2010.

Hubble Space Telescope (HST) is a space-based NASA observatory that has examined the universe in the (mainly) visible portion of the electromagnetic spectrum. This Web site contains contemporary information and educational materials about astronomy, astrophysics, and cosmology, including topics such as black holes, neutron stars, dark matter, and dark energy. Available online. URL: http://hubblesite.org/. Accessed on February 12, 2010.

Institute and Museum of the History of Science in Florence, Italy, offers a special collection of scientific instruments (some viewable online), including those used by Galileo Galilei. Available online. URL: http://www.imss.fi.it/. Accessed on February 12, 2010.

International Union of Pure and Applied Chemistry (IUPAC) is an international nongovernmental organization that fosters worldwide communications in the chemical sciences and in providing a common language for chemistry that unifies the industrial, academic, and public sectors. Available online. URL: http://www.iupac.org/. Accessed on February 12, 2010.

National Aeronautics and Space Administration (NASA) is the civilian space agency of the U.S. government and was created in 1958 by an act of Congress. NASA's overall mission is to direct, plan, and conduct American civilian (including scientific) aeronautical and space activities for peaceful purposes. Available online. URL: http://www.nasa.gov/. Accessed on February 12, 2010.

National Institute of Standards and Technology (NIST) is an agency of the U.S. Department of Commerce that was founded in 1901 as the nation's first federal physical science research laboratory. The NIST Web site includes contemporary information about many areas of science and engineering, including analytical chemistry, atomic and molecular physics, biometrics, chemical and crystal structure, chemical kinetics, chemistry, construction, environmental data, fire, fluids, material properties, physics, and thermodynamics. Available online. URL: http://www.nist.gov/index.html. Accessed on February 12, 2010.

National Oceanic and Atmospheric Administration (NOAA) was established in 1970 as an agency within the U.S. Department of Commerce to ensure the safety of the general public from atmospheric phenomena and to provide the public with an understanding of Earth's environment and resources. Available online. URL: http://www.noaa.gov/. Accessed on February 12, 2010.

NEWTON: Ask a Scientist is an electronic community for science, math, and computer science educators and students sponsored by the Argonne National Laboratory (ANL) and the U.S. Department of Energy's Office of Science Education. This Web site provides access to a fascinating list of questions and answers involving the following disciplines/topics: astronomy, biology, botany, chemistry, computer science, Earth science, engineering, environmental science, general science, materials science, mathematics, molecular biology, physics, veterinary, weather, and zoology. Available online. URL: http://www.newton.dep.anl.gov/archive.htm. Accessed on February 12, 2010.

Nobel Prizes in Chemistry and Physics. This Web site contains an enormous amount of information about all the Nobel Prizes awarded in physics and chemistry, as well as complementary technical information. Available online. URL: http://nobelprize.org/. Accessed on February 12, 2010.

Periodic Table of Elements. An informative online periodic table of the elements maintained by the Chemistry Division of the Department of Energy's Los Alamos National Laboratory (LANL). Available online. URL: http://periodic.lanl.gov/. Accessed on February 12, 2010.

PhET Interactive Simulations is an ongoing effort by the University of Colorado at Boulder (under National Science Foundation sponsorship) to provide a comprehensive collection of simulations to enhance science learning. The major science categories include physics, chemistry, Earth science, and biology. Available online. URL: http://phet.colorado.edu/index.php. Accessed on February 12, 2010.

ScienceNews is the online version of the magazine of the Society for Science and the Public. Provides insights into the latest scientific achievements and discoveries. Especially useful are the categories Atom and Cosmos, Environment, Matter and Energy, Molecules, and Science and Society.

Available online. URL: http://www.sciencenews.org/. Accessed on February 12, 2010.

The Society on Social Implications of Technology (SSIT) of the Institute of Electrical and Electronics Engineers (IEEE) deals with such issues as the environmental, health, and safety implications of technology; engineering ethics; and the social issues related to telecommunications, information technology, and energy. Available online. URL: http://www.ieeessit.org/. Accessed on February 12, 2010.

Spitzer Space Telescope (SST) is a space-based NASA astronomical observatory that observes the universe in the infrared portion of the electromagnetic spectrum. This Web site contains contemporary information and educational materials about astronomy, astrophysics, and cosmology, including the infrared universe, star and planet formation, and infrared radiation. Available online. URL: http://www.spitzer.caltech.edu/. Accessed on February 12, 2010.

Thomas Jefferson National Accelerator Facility (Jefferson Lab) is a U.S. Department of Energy–sponsored laboratory that conducts basic research on the atomic nucleus at the quark level. The Web site includes basic information about the periodic table, particle physics, and quarks. Available online. URL: http://www.jlab.org/. Accessed on February 12, 2010.

United States Geological Survey (USGS) is the agency within the U.S. Department of the Interior that serves the nation by providing reliable scientific information needed to describe and understand Earth, minimize the loss of life and property from natural disasters, and manage water, biological, energy, and mineral resources. The USGS Web site is rich in science information, including the atmosphere and climate, Earth characteristics, ecology and environment, natural hazards, natural resources, oceans and coastlines, environmental issues, geologic processes, hydrologic processes, and water resources. Available online. URL: http://www.usgs.gov/. Accessed on February 12, 2010.

Index

Italic page numbers indicate illustrations.

C